A SONNET TO SCIENCE

MANCHESTER
1824

Manchester University Press

A sonnet to science

Scientists and their poetry

Sam Illingworth

Manchester University Press

Published by Manchester University Press
Altrincham Street, Manchester M1 7JA
www.manchesteruniversitypress.co.uk

British Library Cataloguing-in-Publication Data
A catalogue record for this book is available from the British Library

ISBN 978 1 5261 2798 3 hardback

First published 2019

Typeset in Monotype Fournier by
Servis Filmsetting Ltd, Stockport, Cheshire
Printed in Great Britain by
TJ International Ltd, Padstow

For my wife, Becky

Contents

Acknowledgements

I would like to thank all of the people and organisations who have helped this book to become a reality, starting with the Royal Society, whose grant made it possible for me to conduct the research that went into its writing. I would also like to thank Manchester Metropolitan University, my employer, for granting me the time to both research and write this book, and all of my colleagues there who continue to support and inspire me in my research and teaching.

Many thanks to the various reading rooms and staff at the British Library, the National Library of Scotland, the National Poetry Library, and the Special Collections and Archives at the University of Liverpool, who always made me feel welcome and who provided help, advice, and resources that were essential throughout the process of researching and writing. Thank you to the archivists at Somerville College, Oxford, the Institution of Engineering and Technology, the University of Manchester, and the Bodleian Libraries, who were a great help in finding various manuscripts. Thank you also to the Revd Tom Meyrick who helped me to track down Ada Lovelace's memorial in Kirkby Mallory.

A special thanks is needed for the various readers who provided specialist assistance for several of the chapters: to Sharon Ruston for her help with the Humphry Davy chapter, Mary Gibson for her help with the Ronald Ross chapter, Jan Čulík for his assistance with the Miroslav Holub chapter, and finally Anne Berkeley, Sarah Elson Brug, Petà Dunstan, and Angelo di Cintio for all of the information and testimony that they provided about Rebecca

Elson. Jean-Patrick Connerade provided advice, translations, and inspiration for various sections of this book, and has been a great support throughout. All of these readers have helped to improve the readability and reliability of the material here, but any mistakes that occur remain mine alone.

Thank you to Bloodaxe Books for permission to reproduce the poems and extracts of Miroslav Holub, and also to Carcanet Press for permission to reproduce the poems and extracts of Rebecca Elson. Thanks also to Manchester University Press for the opportunity to write this book; in particular to the anonymous reviewers, whose comments have helped to improve its quality almost immeasurably, and to Tom Dark, who has been a great ally and editor throughout the process. Finally, thank you to my family and friends who support me in everything I do, and who have always encouraged me to keep asking questions. A special thank you to my friend Dan Simpson, without whom I would not have had the confidence to continue to develop my research into the relationship between science and poetry, and to my wife Becky, without whom life would be remarkably unpoetic

Introduction

Science! true daughter of Old Time thou art!
  Who alterest all things with thy peering eyes.
Why preyest thou thus upon the poet's heart,
  Vulture, whose wings are dull realities?
How should he love thee? or how deem thee wise,
  Who wouldst not leave him in his wandering
To seek for treasure in the jewelled skies,
  Albeit he soared with an undaunted wing?
Hast thou not dragged Diana from her car,
  And driven the Hamadryad from the wood
To seek a shelter in some happier star?
  Hast thou not torn the Naiad from her flood,
The Elfin from the green grass, and from me
The summer dream beneath the tamarind tree?
                    'Sonnet – To Science' by Edgar Allan Poe[1]

When he wrote 'Sonnet – To Science' in 1829, Poe was rallying behind the sentiments of John Keats in the following section from his narrative poem 'Lamia', written in 1819:

There was an awful rainbow once in heaven:
We know her woof, her texture; she is given
In the dull catalogue of common things.
Philosophy will clip an Angel's wings,
Conquer all mysteries by rule and line,
Empty the haunted air, and gnomèd mine—
Unweave a rainbow, as it erewhile made
The tender-personed Lamia melt into a shade.[2]

The concern of both Keats and Poe was that science would steal away some of the magic from the world; the rainbow of nature would be unwoven by the technological advancement of science, with the colour of the imagination replaced by monochrome facts. When writing these poems at the beginning of the nineteenth century, the two poets were living during a time of great upheaval and intellectual conflict, Romanticism's celebration of nature and individualism opposing the new technologies and collectivism of the Industrial Revolution. As a writer, it is perhaps no wonder that Poe felt this way, his poetry reflecting the uncertainty of what lay ahead and the worry that there would be no place left for his craft in this new world order. Keats, himself a trained apothecary and surgeon, was arguably better placed to make a more rational judgement, but like Lord Byron and others before him he was wary of the double-edged nature of these technological advances and the negative effect that they might have on both nature and the working classes.

When I first read Poe's poem I found myself disagreeing with his sentiment. Instead, I believe that the more we find out about science, the more we realise what a beautiful and incredible world we live in. The fact that the patterns on a pine cone match those of the Fibonacci sequence, or that we can stare back into the void a few seconds after the creation of the universe, are, to me, beautiful things. Like many fears, I think that those of Poe and others were based in recognition or familiarity, and that in reality, science and poetry are actually very similar. For example, there are many overlaps in the process of writing a poem or conducting a scientific experiment. When you begin experimenting in either discipline, you have to follow rules and regulations that produce half-expected results; it is only by fully exploring these rules that you get an underlying sense of how they can be used to create your own work, or how they must be rejected in favour of a new form or hypothesis.

It was a rejection of Poe's sentiments that led to the development of 'Peer Reviewed Poetry', a spoken word show that I toured across the United Kingdom with my friend and colleague Dan

Simpson.[3] This show involved a somewhat straw man argument centred around which discipline was better, 'science' (defended by myself) or 'poetry' (for which Dan put forward the case), with the discussion focused on a collection of poems that had either been written about science or by scientists themselves. In doing the research for this show it quickly became apparent that a number of well-known scientists had written poetry, and that while the aesthetic quality of their output was something of a mixed bag, it was not immediately apparent *why* they had written it. Did writing poetry help them to make better sense of the world in which they lived? Did they consider themselves to be talented polymaths? Was it purely a private pursuit of pleasure?

It was a desire to better understand this line of enquiry that led me to successfully apply for a research grant from the Royal Society entitled 'How has poetry expanded scientists' understanding of the world in which we live?' This research aimed to investigate which other scientists wrote poetry, and their motivation for doing so. Taking inspiration from the Austrian philosopher Ludwig Wittgenstein and his maxim that 'the limits of my language mean the limits of my world',[4] I wanted to determine whether scientists who embraced poetry were also increasing their understanding of the world, expanding their language and thereby their capacity to communicate their science to others. By reading the poetry of scientists and contextualising their verse alongside their scientific endeavours, I aimed to establish the *why*, and in doing so better understand the effectiveness of such an approach for science communication. This book is a result of that research, and while some might argue that such a scientific approach to poetry is incongruous, it is arguably the path through which both Western science and literary criticism arose, courtesy of Aristotle.

Aristotle is considered to be one of the greatest philosophers of all time; in addition to being the forefather of Western science, his *Poetics*[5] is thought by many academics to be the starting point for literary theory in Western culture.[6] In this treatise, Aristotle approaches poetry with the same methodology that he had previously adopted when discussing biology and physics in texts such as *Physics* and *The History of Animals*.[7] By collecting data in the form of poetry, analysing it, and discussing the implications of his analysis, Aristotle proposed a number of distinct classifications and conclusions; for example, the notion that *mythos* (or plot) is the most important and dominant aspect of a tragic poem. This approach to poetic analysis will be familiar to any scientists who are reading this book, as it is effectively the scientific method: propose a hypothesis, gather data, make observations, analyse these observations, and then revisit the original hypothesis to either accept, amend, or reject it.

While scholars accept that there is some merit to this scientific approach to literary analysis, they also largely agree that poetry cannot be studied in the same way as science.[8] They argue that while the scientific method is dependent on a set of underlying laws and characteristics that govern the behaviour of the natural world, poetry, and art in general, is often driven by questioning the accepted assumptions of previous generations. However, I would argue that as with poetry, science also thrives by questioning the conventions and norms that a previous generation has accepted. As such, it is perhaps naïve to assume that poetry cannot also be characterised using a systematic approach more readily adopted by the scientific method. Likewise, more modern forms of literary criticism can also be applied to the fields of frontier science, and indeed there exist several studies that try to bring together these two seemingly dissimilar analytical approaches, with examples including a quantum approach to literary deconstruction and feminist explorations of gerontology.[9]

Aristotle himself points to the differences between the two disciplines, arguing in *Poetics* that while poetry paints an imaginative picture, physical philosophy (i.e. science) deals only in facts. However, once more I find myself disagreeing with this viewpoint, as poetry can deal in facts just as science can be used to paint imaginative pictures. To demonstrate this argument, consider the following two extracts, both of which concern the yellowhammer, a sparrow-sized, bright yellow bird that is native to Eurasia and found across many parts of the UK. The first is taken from

4

the methods section of a scientific research article written in 2000, entitled 'Habitat associations and breeding success of yellowhammers on lowland farmland', and the second is an extract from 'The Yellowhammer's Nest', a poem written by the poet John Clare between 1824 and 1835:

> For nests that failed, the date and, where possible, the cause of failure were recorded (e.g. starvation, predation, nest collapse). The date of failure (or fledging) was estimated as the mid-point between the date when the nest was last known to be active and the date on which it was found to have failed (or the fledglings to have left the nest). When nestling age was not known precisely from observation of hatching, it could be estimated by comparing the degree of feather development of the largest nestling with known-age broods.[10]

> Yet in the sweetest places cometh ill
> A noisome weed that burthens every soil
> For snakes are known with chill and deadly coil
> To watch such nests and seize the helpless young
> And like as though the plague became a guest
> Leaving a housless-home, a ruined nest
> And mournful hath the little warblers sung
> When such like woes hath rent its little breast[11]

Both of these passages are concerned with describing the 'site' of a yellowhammer's nest, and the risk of predation and starvation that the occupants face. The poetry of Clare is factually accurate, as indeed is much of his writing on birds and nature; for example, earlier in the poem he describes the contents of the nest as: 'Five eggs pen-scribbled o'er with ink their shells / Resembling writing scrawls which fancy reads'. The folk-name for the yellowhammer is the 'scribble-lark', due to the mesh of fine dark lines that cover their eggs and make them look as though someone has scribbled over them with a pen.[12] This is one of several facts that are contained within Clare's poem. Likewise, while the language used in the scientific research paper is somewhat academic, words and phrases such as 'nest collapse', 'starvation', and 'degree of feather development' cannot be read without conjuring up specific imagery; a narrative relating to the plight of the yellowhammers under observation quickly forms in one's mind. The words of the poet and the scientists thus both present an argument that deals in facts, while also painting imaginative pictures.

Were Poe and Aristotle both incorrect in their separation of these two disciplines? Could poetry and science cohabit the same spheres of fact and imagination, and in doing so offer us a more complete image of the world and an understanding of how it works? By choosing scientists who also wrote poetry, I hoped to be able to examine more closely the relationship between science and poetry, in the context of individuals who had a personal commitment to both disciplines.

So where to start? There are already a small number of high-quality books that examine the nature of science and poetry. *Science and Poetry* by Mary Midgley is a mainly philosophical text that challenges the concept that science rather than poetry has a 'right' to explain how the universe operates.[13] While some of the ideas that are presented in Midgley's book underpin the ethos of *A Sonnet to Science*, the context of the two books is different, with the former not primarily focused on individual scientists. Similarly, *Science in Modern Poetry: New Directions* is a collection of essays edited by John Holmes that is mainly concerned with how science has influenced modern poets, and not how poetry has influenced scientists.[14] Peter Middleton's *Physics Envy: American Poetry and Science in the Cold War and After* presents a broader history of how science and poetry have worked together to find universal truths, examining the work of poets and scientists such as Oppenheimer and Heisenberg.[15] However, it is limited to the period of history surrounding the Cold War, and is more concerned about the sociological influence behind the poetry of the time rather than why certain scientists wrote poetry and the relationship between their writing and their scientific research. *The Poetry of Victorian Scientists: Style, Science and Nonsense* by Daniel Brown presents an overview of the lives of several Victorian scientists who wrote and were influenced by poetry, and perhaps most closely matches the aims of my own research; however, while it is an excellent book it again deals with a very specific time period, and is perhaps aimed at a core audience of literary scholars.[16] While I hope that *A Sonnet to Science* will similarly appeal to this audience, I wanted to write a book that would also appeal, and be easily read, by a more general audience. *Contemporary Poetry and Contemporary Science* presents a collection of essays written by both contemporary poets and scientists, and features an afterword by Gillian Beer, who has also written widely on the relationship between literature and science.[17] However, this book investigates the similarities and differences in the way that poets and scientists examine the world around them, but for the most part it is concerned with the opinion of either scientists or poets, rather than being an exploration of scientists who wrote poetry and the effect that this had on their research and practice.[18]

By highlighting the work of several scientists and the role that poetry played in both their personal lives and their scientific achievements I wanted *A Sonnet to Science* to present an aspirational account of how the two disciplines can work together, and in doing so hopefully convince current and future generations of scientists and poets that these worlds are not mutually exclusive, but rather complementary in nature. At this point I must also make a confession and declare that this goal is not entirely altruistic, since it is also driven by my own identity as a scientist and a poet, and my struggle to better understand the relationships between the two disciplines. The research that I conducted for 'Peer Reviewed Poetry' revealed a large number of scientists who wrote poetry, and in selecting the scientists to focus on for this book, the following three criteria were applied: the scientists had to be Western; their poetry had to be readily available in English, with any translations approved by the poet; and the poets had to occupy a continuous timeline from around the end of the eighteenth century until the present day.

It was first necessary for me to limit the scientists that feature in this book to those from the Western world. This was because the book is grounded in a Western approach to the perception and dissonance of science and poetry, as set forth by both Aristotle and Poe. Indeed, the relationship between science and poetry is far less fractured in many other regions of the world than it is in the West. This meant that luminaries such as Ahmed Zaki Abu Shadi, the Egyptian Romantic poet and bacteriologist, and A. P. J. Abdul Kalam, the space scientist, poet, and former president of India, would be omitted. I accept that this is a failing of the aspirational ambitions of this book, especially in light of the issues of diversity that face STEM (Science, Technology, Engineering, and Maths) researchers working and studying in the West, but I hope that it can serve as an impetus for future studies.[19]

One of my many personal failings is that I am only fluent in one language, English; as such I am only able to comment on and analyse poetry that is written in English. Where translations do exist for non-English poems I only considered those that were sanctioned by the author, both as a mark of respect and to further ensure the validity of any analysis. This meant that several important scientific and literary figures were excluded from the selection, most notably the Russian polymath Mikhail Vasilyevich Lomonosov, for whom author-sanctioned translations do not, to the best of my knowledge, exist. It is important to acknowledge that for one of the scientists in this book, Miroslav Holub, there are some reported issues with his translations from Czech into English, but the poet himself was satisfied with those that are used in this book.[20]

In determining the criteria for the timeline, I thought it was essential to ensure that there was an overlap between each of the scientists, both in terms of their lifespans and their activity as researchers, so that a narrative of the developing relationship between the disciplines could be established. Given that this book came about as a direct response to a poem written by Edgar Allan Poe at the beginning of the nineteenth century, I did not want to extend too far beyond this timeframe. Arguably the most important person to be excluded because of this criterion is the English physiologist and botanist Erasmus Darwin, but I would urge interested readers to seek out Martin Priestman's *The Poetry of*

Erasmus Darwin: Enlightened Spaces, Romantic Times, which presents a brilliant account of Darwin's literary accomplishments and scientific achievements.[21] Choosing the cut-off date at around the beginning of the nineteenth century was also important because this is when the term 'scientist' first came into being; it was suggested by the English polymath William Whewell at a meeting of the British Association for the Advancement of Sciences on 24 June 1833 as an analogy to artist. Interestingly, this suggestion was put forward in response to a criticism by the Romantic poet Samuel Taylor Coleridge (who will feature later in this book), who believed that the term philosopher, while applying to his own achievements as a writer, did not apply to the majority of the association's members. This coining of the phrase 'scientist' heralded the creation and division of a large number of scientific societies, which drove forward the professionalisation of science, and was in part responsible for further divisions between poetry and science. This new professionalism meant that my selection criteria could not reasonably have included any 'poets who practised science', as science became a far more specialised and expensive pursuit.

Ultimately, I settled on the following scientists, whose selection perfectly matched the criteria that I had set out: Humphry Davy (1778–1829), Ada Lovelace (1815–52), James Clerk Maxwell (1831–79), Ronald Ross (1857–1932), Miroslav Holub (1923–98), and Rebecca Elson (1960–99). Despite the arguments presented above, I know that there are many scientist-poets whom I have missed out, both those with several poetry collections (such as Edward Lowbury and Roald Hoffmann), and those who wrote less regularly, but for whom poetry had a profound effect on their work (for example, Julius Robert Oppenheimer). However, this is a personal selection, albeit one that is informed by a methodological approach, not dissimilar to that applied by Aristotle in his *Poetics*. By presenting the poetic accomplishments of these scientists alongside their life histories and scientific achievements I hope that I am able to provide a tentative explanation as to why these scientists wrote poetry, and how their individual stylings and reasons for writing verse might have concurred. In doing so I hope to demonstrate that the two disciplines offer complementary, rather than antagonistic, viewpoints for understanding the world and the way in which we live.

1

The Romantic scientist: Humphry Davy

Theirs is the glory of a lasting name,
  The meed of genius, and her living fire;
Theirs is the laurel of eternal fame,
  And theirs the sweetness of the muse's lyre.
  From 'Sons of Genius' by Humphry Davy[1]

Good master Davy

Humphry Davy was born on 17 December 1778 in Penzance, Cornwall to Grace (née Millet) and Robert Davy. Robert had initially trained as a wood carver, although the death of his own father resulted in an inheritance of property that meant that he was able to pursue this trade as an artful hobby rather than as a full-time job. Grace had been orphaned at an early age, her parents both killed by a lethal fever, and her mother's dying wish had been for her friend, the eminent local surgeon and apothecary John Tonkin, to adopt Grace and her sisters. Tonkin, who was on hand during the death of Grace's mother, agreed and took the three siblings into his home, looking after them until they were married.

Davy was the eldest of five children, and at the age of five was placed in a preparatory seminary,[2] where the headmaster quickly noted his intellect and recommended transfer to a school that would enable a superior education. At the age of six, Davy was therefore placed in the Latin School in Penzance (the precursor to the Penzance Grammar School), remaining there even after his family moved to their estate at Varfell, a hamlet in the nearby

11

parish of Ludgvan. It was roughly three miles between the two locations, a distance that was considered too far for the young Davy to travel on a daily basis, and so during term time he lived with his adopted grandparent John Tonkin, returning to his parents at Varfell during the holidays. Tonkin also bought Davy a horse, and after teaching him how to ride it the two of them would ride every Saturday from Penzance to the family estate.

At that time the Latin School was under the care of the Revd George Coryton, a man who has since been described as being 'ill-fitted for the office of teaching'.[3] Coryton was renowned for corporal punishment, and composed the following punitive piece of doggerel especially for Davy, which he would recite while repeatedly hitting the young boy with a ruler across his open palms:[4]

> Now, Master Dàvy,
> Now, Sir, I hàve 'e,
> No one shall sàve 'e,
> Good Master Dàvy!

As we shall see later, this was not the only piece of verse that utilised the assonance of Davy's name to cause him pain and injustice.

By the age of 14 Davy had finished the local curriculum and so moved from the Latin School to finish his education at Truro Grammar School, under the tutelage of the Revd Cornelius Cardew. Cardew also had a reputation for severity, and would later remark that he did not remember Davy being a particularly outstanding student, with any talent that he did show being in his translations of the classics rather than any scientific pursuits. However, it is clear that Davy was equally nonplussed about the schooling that he received in both Penzance and Truro; writing to his mother in 1802 about his brother John's schooling, he notes:

> I recollect I was rejoiced when I first went to Truro school; but I was much more rejoiced when I left it for ever. Learning naturally is a true pleasure how unfortunate then it is that in most schools

12

it is made a pain. And yet Dr Cardew comparatively was a most excellent master. I wish John may have as good a one. — After all the way in which We are taught Latin & Greek does not much influence the important structure of our minds. I consider it as fortunate that I was left much to myself when a Child & put upon no particular plan of study & that I enjoyed much idleness at Mr Coryton's school. I perhaps owe to these circumstances the little talents I have & their peculiar application. What I am I have made myself, I say this without vanity & in pure simplicity of heart.[5]

In these early years, Davy enjoyed spending time reading, sketching, fishing (a pastime that he would continue to pursue throughout his life), and regaling his schoolmates with made-up tales of adventure. During this time, Davy also began experimenting with chemistry, including the manufacture of basic fireworks and the construction of a lantern from a turnip and a candle that he then used to melt tin, charging friends who wished to observe the spectacle. As well as being in demand for his early experiments and storytelling, the young Davy, aided by his translation skills, was also apparently sought out as a creator of poetry and valentines, demonstrating at this early age his ability to awe through either lyrical verse or scientific experimentation.

After a year of boarding at Truro Grammar School, Davy, having now finished his formal schooling, returned to Penzance in December 1793, where he once more took up residence with Tonkin. Over the next year Davy learned French from a refugee priest who was fleeing the French Revolution, and one of his earliest surviving poems comes from around this period, written in French and addressed to a Frenchwoman living in Penzance whom Davy was trying to woo; it begins:[6]

L'Etre infini dont l'image réside
Dans le ciel bleu,
Dans les splendeurs de l'Ocean limpide,
Règne en tout lieu.
[The Supreme Being whose image resides
In the infinite blue sky,

In the splendours of the limpid Ocean,
Rules over all.][7]

Other than learning French, Davy appears to have done little of note during this time, with his brother John remarking in his biography that:

> The greater part of the following year he was, I believe, in an unsettled state, studying in a desultory manner, by fits and starts, and yielding to the allurements of occasional dissipation, and the amusements which constitute the delight of active youth, as fishing, shooting, swimming and solitary rambles.[8]

However, these rambles and explorations of nature were to have a profound effect on the literary stylings of Davy, influencing both his poetry and philosophical thinking. During this year of idleness, at the end of 1794 and just a few days before his sixteenth birthday, Davy's father succumbed to a fatal infection. Now that he was the most senior man of the house, Davy was expected to pursue a career to support his family; once more he was indebted to Tonkin, who set him up as an apprentice to John Bingham Borlase, a local surgeon with a practice in Penzance.

With the commencement of his apprenticeship, Davy also began making careful records in a notebook. In addition to notebooks that were kept for entirely scientific reasons, Davy also kept many personal notebooks, in which he wrote over 160 poems and fragmentary lines, throughout the course of his lifetime. In all his preserved notebooks (most of which are archived at the Royal Institution in the UK), Davy wrote from both directions, and in his personal notebooks he regularly mixed scientific observations with poetic experimentation. The first of his notebooks commences with a plan of study, outlining the topics that Davy planned to study during his apprenticeship, and the order in which he sought to do this. This study plan was impressive, beginning with theology and geography before moving on to anatomy, surgery, and chemistry (all of which were skills that Davy identified as key to his profession as an apprentice surgeon), while also encompassing physics, mathematics, history, and seven languages.

The cover of this first notebook displays two images: on one side is an ancient lyre, and on the other is an olive wreath encircling a lamp. These images would seem to suggest that even at this early stage of his career, Davy considered the poetic (as represented by the lyre – a Greek stringed instrument that is associated with ballads and poetry) to be a complement to the scientific (as represented by the lamp), and given the significance that the mining lamp would later play in his life, this is a fitting and somewhat prophetic image.[9] Alongside his plans for study, Davy also laid out a prospectus for a future volume of poetry; an ambitious collection of eight odes, four poems dedicated to Cornish scenes, and a longer ballad called 'The Lady'. Davy eventually succeeded in writing many of these poems, the best-known of which is 'Sons of Genius', written in 1795, which begins:

> Bright bursting through the awful veil of night
>   The lunar beams upon the ocean play;
> The watery billows shine with trembling light,
>   Where the shift breezes skim along the sea.

> The glimmering stars in yon ethereal plain
>   Grow pale, and fade before the lurid beams,
> Save where fair Venus, shining o'er main,
>   Conspicuous still with fainter radiance gleams.[10]

This poem consists of 32 stanzas, and discusses how a select group of men possessed with genius are immortalised by their achievements; the final stanza reads as follows:

> Theirs is the glory of a lasting name,
>   The meed of genius, and her living fire;
> Theirs is the laurel of eternal fame,
>   And theirs the sweetness of the muse's lyre.

This final stanza explicitly refers to the two images that adorn Davy's earliest notebook: the lyre and the olive wreath (laurel) encircling the lamp. In analysing this poem, it could be argued that its content is indicative of Davy's immediate inspiration from nature and the surrounding Cornish landscape, and how these landscapes inspired him to write poetry that grappled with the notion of human existence.[11] The poem is also representative of Davy's ambitions, and by comparing himself to luminaries such as Homer and Newton he is not only marking himself out for future greatness, but in a similar vein to his study plan and poetry prospectus, he is creating a road map for how he will achieve the same immortality as the sons of genius that he writes about. The fascination with fame and recognition that resonates throughout the verses of 'Sons of Genius' is something that would play a prominent part in Davy's poetic and scientific development.

At the end of 1797 Davy's plan of study eventually saw him move on to the topic of chemistry, with his initial reading on the subject confined to two texts: Antoine Lavoisier's *Elements of Chemistry*, translated and published in 1790, and William Nicholson's *Dictionary of Chemistry*, published in 1795. While Nicholson's text provided a guidebook for how chemistry could be *done*, Lavoisier (considered by many to be the founder of modern chemistry) showed what chemistry could *achieve*. It was chemistry, and Lavoisier in particular, that first provoked in Davy a need not only to absorb information and knowledge, but also to question and advance the work of others. After only a few months of study Davy had developed a new hypothesis to rival some of Lavoisier's work, and he set about devising experiments to prove his theories, using a room in Tonkin's house as his makeshift laboratory, and his elder sister Kitty as his apprentice. At around this time, Davy was also seeking inspiration from a local tradesman called Robert Dunkin. A saddler by trade, Dunkin was also well known for his interest in science and engineering, and was regarded by many as an able mathematician, highly skilled in the areas of electricity and magnetism. Dunkin helped Davy to sound out many of his ideas, assisting Davy in creating the equipment that he needed to test his theories, and providing an audience for these early experiments. One of the most significant of these experiments was when Davy demonstrated the immateriality of heat by rubbing two blocks of ice against one another, causing them to melt together, and thereby demonstrating how heat could be generated by friction, an idea that was contrary to Lavoisier's 'caloric theory', which stated that ice would only melt if put into contact with a hotter body, thereby releasing the flow of 'caloric particles' into the ice and causing it to melt. These experiments with Dunkin demonstrated to Davy that, like poetry, experiments could benefit from being shared and appreciated by an audience. Just as he had done with the storytelling of his youth, Davy now sought new audiences to regale with his scientific prowess.

In seeking out these new audiences, Davy was introduced to Davies Giddy, a Cornish engineer and politician. Giddy met Davy through his friend John Denis, a local banker whom Davy had originally tried to impress with his experiments, but who (having no knowledge of such things) had instead asked Davy if he would like to meet someone who better understood these topics. After showing Giddy his work on the immateriality of heat, Davy was encouraged to send his research to Thomas Beddoes, the English physician and chemist. Beddoes had been the Professor of Chemistry at Oxford University but had resigned from his post in 1792, moving to Bristol where he set up a clinic to investigate treatments for tuberculosis. This later became the Pneumatic Institute, an independent centre of research which aimed to determine if the inhalation of various gases might present a potential cure for tuberculosis. After reading Davy's treatise on the immateriality of heat, and at the recommendation of his good friend Giddy, Beddoes offered the role of superintendent to Davy. There then followed a period of negotiation between Giddy, Beddoes, Tonkin, and Davy's mother, Grace. Tonkin and Grace were both keen for Davy to continue his apprenticeship under Borlase, but after receiving assurances that he would continue to train as a physician, they eventually acquiesced. Furthermore, due to the generous terms of his employment, Davy could re-assign all rights to his father's inheritance to his mother, thereby helping to ensure the livelihood of the Davy family. And so, on 2 October 1798, aged just 19, Davy set off to Bristol to begin work as the superintendent of the Pneumatic Institution.

Clad with new born mightiness

When Davy arrived at the Pneumatic Institute, he was still unaware of several major bodies of work and recent research in both chemistry and the wider sciences. However, Beddoes saw in him the potential to achieve great things, remarking to Erasmus Darwin that he found Davy to be: 'The most extraordinary person that I have ever seen, for compass, originality and quickness of thought.'[12] Beddoes was clearly very impressed with Davy, not only in terms of his scientific prowess but also his literary skills. Soon after his arrival in Bristol, Davy was introduced to the Clifton literary circle via the well-connected Beddoes family, who were close friends with, among others, Samuel Taylor Coleridge and Robert Southey, two poets who would play an important role in the development of poetry during the Romantic era. These poets admired and respected Davy, not only as a brilliant chemist, but also as a gifted poet. Indeed, when Southey published the first volume of the *Annual Anthology* in 1799 it contained five of Davy's poems, with the 1800 volume containing a further poem from Davy; several of these poems were those that Davy had originally set out to write in his Portfolio (including 'Sons of Genius').[13] Beddoes also contributed three poems to the *Annual Anthology*, one of which was a parody of Erasmus Darwin's 'Botanic Garden'.

In addition to publishing alongside his new literary friends, Davy also found a way of involving them in his experiments. As part of his research into the medical powers of experimentally produced gases, Davy conducted several studies involving nitrous oxide, using a chamber that had been specially built for the Pneumatic Institute by the Scottish inventor and chemist James Watt. Nitrous oxide is a colourless and almost odourless gas that was first discovered in 1772 by the English scientist and clergyman Joseph Priestley, and Beddoes was convinced that the inhalation of the gas would lead to medical benefits. These beliefs were in spite of earlier research conducted by the American physician Samuel L. Mitchill, which stated that nitrous oxide would probably be fatal if inhaled. Davy was tasked with investigating Beddoes' theory, coercing his friends into becoming guinea pigs, and asking them to comment on the effects that inhaling the gas had on their mental and physical capabilities. As well as observing that nitrous oxide could potentially relieve a person from pain (something which would not be utilised for several decades more, when nitrous oxide was adopted as an anaesthetic in dentistry), Davy quickly realised the euphoric effects of the compound, leading him to give the gas its alternative name, laughing gas, and sparking the use of nitrous oxide as a recreational drug among the upper class.

During these experiments, Davy seems to have become addicted to nitrous oxide, and would often go back into the laboratory late at night to make further observations. However, throughout these observations he continued to make detailed notes about the effect that the gas was having on both his mental and physical states. Davy recorded some of these early experiments in verse, partly to determine the extent to which his linguistic skills were affected, and partly to explore whether what he was experiencing could be imaginatively described; 'On breathing the Nitrous Oxide' was written during this period:[14]

> Not in the ideal dreams of wild desire
> Have I beheld a rapture wakening form
> My bosom burns with no unhallowed fire
> Yet is my cheek with rosy blushes warm
> Yet are my eyes with sparkling lustre filled
> Yet is my mouth implete with murmuring sound
> Yet are my limbs with inward transports thrill'd
> And clad with new born mightiness round —

This poem can be thought of as a piece of scientific data, which describes with quite precise physiological information the effects of Davy's inebriation, while offering a personal and intimate account of his exposure to the gas; a different perspective to the research that was later published on the topic. Despite his own efforts, Davy was unsuccessful in convincing Coleridge and the other poets to write about the effects of laughing gas while intoxicated, though Coleridge himself would later write one of his most famous poems, 'Kubla Khan', while under the influence of opium,[15] thereby mirroring the behaviour of Davy, although to significantly greater critical acclaim ('On breathing the Nitrous Oxide' was not published during Davy's lifetime).

Despite being unable to convince his friends to compose poetry while under the influence of nitrous oxide, Davy's own poetic abilities were held in high esteem. Indeed, following his relocation to the Lake District in 1800, and after a recommendation from Southey and Coleridge, the great Romantic poet William Wordsworth asked Davy to proofread the second edition of *Lyrical Ballads*, the first manuscript to include his famous 'Preface', which laid many of the foundations for the Romantic era's revolt against the constraints of Augustan poetry. Wordsworth's respect for Davy can be seen in this extract from a letter that he wrote to him in July 1800:

> You would greatly oblige me by looking over the enclosed poems and correcting anything you find amiss in the punctuation a business at which I am ashamed to say I am no adept ... I write to request that you would have the goodness to look over the proofsheets of the 2nd volume before they are finally struck off.[16]

This was a great responsibility, and in editing and proofreading the poems in *Lyrical Ballads*, Davy was probably influenced in his future thinking as both a scientist and a poet; the following poem is taken from one of Davy's notebooks, in which he adopts the ballad form of Wordsworth:

> As I was walking up the street
> In pleasant Burney town
> In the high road I chanced to meet
> My cousin Matthew Brown

My cousin was a simple man
A simple man was He
His face was of the hue of tan
And sparkling was his eye. –

His coat was red for in his Youth
A soldier He had been.
But He was wounded & with ruth
He left the camp I ween –

His wound was cured by Dr John[17]
Who lives upon the hill
Close by the rock of grey free stone
And just above the rill. –

He then became a farmer true
Who took to him for aid
A wench who though her eye was blue
Was yet a virgin maid. –

He married her & had a son
Who died in early times
And in the churchyard is made Known
By poet Wordsworths Rhymes.

As long as this fair wife did prove
To him a wife most true
His red coat He away did shove
And wore a coat sky blue.[18]

As is evident from his recommendation to Wordsworth, Coleridge rated Davy's qualities as a poet extremely highly, later observing that 'If Davy had not been the first Chemist, he would have been the first Poet of his age.'[19] Coleridge also wrote to Davy in October 1800 with a detailed critique of one of his poems, 'Written after Recovery from a Dangerous Illness', offering advice for someone that Coleridge considered to be an equal. This poem had originally been entitled 'The Spinosist', and it was a poem that

Davy would revisit and re-edit throughout the rest of his life. It was written as a consideration of the work of the Dutch philosopher Baruch Spinoza, who proposed in the middle of the seventeenth century that God and nature were synonymous, that the whole of the universe (including humans) are made of this nature, and that when we die we return to it. Another core element of this philosophy is that the universe is deterministic, that is, everything that has happened or will ever happen could not have occurred in any other way.

The different variations of this poem, which occur throughout Davy's notebooks, contain some beautiful imagery, and represent a development of Davy's personal philosophy that all human bodies have an immortal soul that is sealed in the body until death, at which point it is released and returned into nature. This worldview, as well as being in line with that of some of the Romantic poets,[20] is also highly reflective of the science that Davy was championing: that chemical properties do not solely rely on their material constituents, but instead depend on the way that they react with other materials (i.e. chemical reactions), as demonstrated by his earlier experiments with the immateriality of ice. 'Written after Recovery from a Dangerous Illness' contains 18 verses, and begins:

Lo! o'er the earth the kindling spirits pour
  The flames of life that bounteous Nature gives;
The limpid dew becomes the rosy flower;
  The insensate dust awakes, and moves, and lives.

All speaks of change: the renovated forms
  Of long-forgotten things arise again;
The light of suns, the breath of angry storms,
  The everlasting motions of the main.

These are but engines of the Eternal will,
  The One Intelligence, whose potent sway
Has ever acted, and is acting still,
  Whose stars, and worlds, and systems, all obey.

Without whose power, the whole of mortal things
    Were dull, inert, an unharmonious band;
Silent as are the harp's untuned strings
    Without the touches of the Poet's hand.

A sacred spark, created by His breath,
    The immortal mind of man his image bears;
A spirit living midst the form of death,
    Oppress'd but not subdued by mortal cares —[21]

In these verses, Davy outlines the deterministic nature of our existence ('The One Intelligence, whose potent sway / Has ever acted, and is acting still') and also the relationship that all living things have with each other ('All speaks of change: the renovated forms / Of long-forgotten things arise again'), while the last two verses make clear Davy's belief in an afterlife, when upon death the soul returns to nature:

Then, as awakening from a dream of pain,
    With joy its mortal feelings to resign;
Yet all its living essence to retain,
    The undying energy of strength divine!

To quit the burdens of its earthly days,
    To give to Nature all her borrow'd powers —
Etherial fire to feed the solar rays,
    Etherial dew to glad the earth with showers.

Southey was also convinced that his friend should have further pursued his poetry, remarking that 'Davy was a most extraordinary man. He would have excelled in any department of art or science to which he had directed the powers of his mind.'[22] Yet despite the opinions of Southey and Coleridge, Davy instead decided to pursue his career as a chemist, acknowledging this commitment himself in a letter to his mother, dated 1 September 1799:

I have sent you with this some copies of a poem on the place of my nativity — but do not suppose I am turned poet. Philosophy,

Chemistry, & medicine are my profession. I had often praised Mount's Bay to my friends here, they desired me to describe it poetically. Hence this poem – which as they admired I published.[23]

In 1799, the same year that the *Annual Anthology* was released, Davy also saw his first scientific publications, with the periodical *Contributions to Physical and Medical Knowledge, Principally from the West of England* dominated by his writings, including 'Experimental essays on heat, light, and on the combinations of light, with a new theory of respiration, and observations on the chemistry of life'.[24] This publication had been set up by Beddoes as a way of allowing his young protégé to demonstrate his scientific prowess to a wider audience, and despite several inconsistencies that would be expected of the still inexperienced author, it was well received, especially among the large and influential group of scientists to which Beddoes was well connected. This group included the English physician Peter Mark Roget[25] and the Scottish physician and chemist Thomas Charles Hope, who was particularly impressed by Davy's work, recommending him to Count Rumford (Benjamin Thompson)[26] for a position at the newly established Royal Institution. Founded in 1799, the Royal Institution had originally been set up to teach and inform the public about new advances in science and technology, with Rumford employing Thomas Garnett, the English physician and philosopher, as a lecturer. However, Garnett's style of delivery proved too dull for the non-scientists who largely made up his audiences, meaning that Rumford needed to employ someone who was a better communicator. Despite Davy having little experience of formal lecturing to a public audience, Hope (and others) strongly recommended him to Rumford, as acknowledged by Davy in this letter to Hope, dated 28 June 1801:

> I believe it is in a great measure owing to your kind mention of me to Count Rumford, that I occupy my present situation in the Royal Institution. I ought to be very thankful to you; for most of my wishes through life are accomplished, as I am enabled to pursue my favourite study, and at the same time to be of some little utility to Society.

This quotation is also notable for the intentions that Davy expresses regarding his new position, and for his categorical statement that it is the pursuit of science (and not poetry) to which he considers himself to be devoted. On 11 March 1801 Davy arrived in London, ready to take up his joint positions as Assistant Lecturer in Chemistry, Director of the Chemical Laboratory, and Assistant Editor of Journals at the Royal Institution.

The lecturer and the lamp

Davy's initial lectures at the Royal Institution were well received, with the young lecturer displaying the same flair for showmanship that he had demonstrated in his youth when burning tin in turnips and melting together slabs of ice with friction. In a letter dated 22 June 1801 to John King, a Swiss surgeon with whom he had worked in Bristol, Davy wrote:

> The voice of fame is still murmuring in my ears – My mind has been excited by the unexpected plaudits of the multitude – I dream of greatness & utility – I dream of Science restoring to Nature what Luxury, what civilization have stolen from her – pure hearts, the forms of angels, bosoms beautiful; & panting with Joy & hope – My labours are finished for the season as to public experimenting & public enunciation My last lecture was on Saturday evening. Nearly 500 persons attended – & amongst other philosophers your countryman Professor Pictet – there was Respiration – nitrous oxide: & unbounded applause. Amen. Tomorrow a party of philosophers meet at the institution to inhale the joy inspiring gas – It has produced a great sensation. Dr Garnett has resigned & at this moment I am the only Lecturer. I hope they will get some professor of mechanics. Before he resigned I was made sole Lecturer on Chemistry. I have been nobly treated by the managers, God bless us I am about 1.000.000 times as much a being of my own volition as at Bristol. My time is too much at my own disposal – So much for egotism – for weak glorious, pitiful, sublime, conceited egotism.

As indicated in this latter, Garnett had by now resigned from his position at the Royal Institution, citing ill health, and opening the door for Davy to became a full lecturer. This letter also high-lights Davy's craving for fame, and while he is somewhat self-deprecating in the acknowledgement of his egotism, there is a distinct lack of empathy regarding the plight of Garnett. Alongside this narcissism, however, there is also a genuinely altruistic goal – Davy wants to demonstrate to the masses how science can help benefit and potentially advance society. Yes, he dreams of the great personal acclaim that will come with this, but he also recognises the societal need for such an approach.

While Davy's lectures at the Royal Institution were influenced by his poetry and poetic friends in both their content and delivery, these lectures in turn influenced the Romantic poets. Coleridge in particular was said to have regularly attended Davy's lectures to improve his stock of metaphors, and he even contacted Davy about setting up a chemistry laboratory for him and Wordsworth to conduct their own experiments; however, due to his other numerous commitments, Davy was unwilling to fully commit to the idea. The insight that Davy presented to his audience of a future civilisation which benefited from the achievements of science clearly influenced Wordsworth, who as well as asking Davy to edit his manuscript for the second edition of *Lyrical Ballads* was also known to have read several of Davy's scientific works. In the third edition of *Lyrical Ballads*, published in 1802, Wordsworth includes a large section about the 'Man of Science' in the preface, part of which reads:

> The remotest discoveries of the Chemist, the Botanist, or Mineralogist, will be as proper objects of the Poet's art as any upon which it can be employed, if the time should ever come when these things shall be familiar to us, and the relations under which they are contemplated by the followers of these respective Sciences shall be manifestly and palpably material to us as enjoying and suffering beings. If the time should ever come when what is now called Science, thus familiarized to men, shall be ready to put on, as it were, a form of flesh and blood, the Poet will lend his divine spirit to aid the transfiguration, and will welcome the Being thus produced, as a dear and genuine inmate of the household of man.[27]

Despite the protestations against science and technology (and in particular the railway) that occur in his poetry, in this passage Wordsworth makes it very clear that science has an important and positive role to play in the development of humankind. That Wordsworth was willing to acknowledge and encourage this role was certainly influenced by his opinion and respect of Davy.

Davy also had a positive effect on other Romantic poets, such as Anna Laetitia Barbauld, who as well as participating in Davy's nitrous oxide experiments, includes a reference to Davy in 'Eighteen Hundred and Eleven: a Poem';[28] and Eleanor Anne Porden, whose long scientific poem *The Veil* contains the following tribute to Davy in the preface:

> The author, who considers herself a pupil of the Royal Institution, being at that time attending the Lectures given in Albemarle-Street, on Chemistry, Geology, Natural History, and Botany, by Sir Humphry Davy, Mr. Brand, Dr. Roget, Sir James Edward Smith, and other eminent men ... was induced to combine these subjects with her story; and though her knowledge of them was in a great measure orally acquired, and therefore cannot pretend to be extensive or profound, yet, as it was derived from the best teachers, she hopes it will seldom be found incorrect.[29]

Davy's appointment as a lecturer did not just result in rave reviews from poets and other members of the public about the accessibility and positive potential for science, it also resulted in a series of scientific discoveries that were to have a significant effect on the development of modern chemistry as we know it today. Davy's main achievements while at the Royal Institution centred on the discovery of six new elements (sodium, potassium, magnesium, calcium, barium, and strontium), which he isolated by using electrolysis, a process by which a liquid or solution containing ions is broken down into simpler substances by passing an electric current through it. For example, Davy isolated sodium by passing an electric current across molten sodium hydroxide, which has the chemical formula $NaOH$ (Na = sodium, O = oxygen, H = hydrogen). In using electrolysis to isolate these substances Davy initiated the field of electrochemistry, in which the effect of electricity on chemical reactions is observed and measured.

Despite moving on from the nitrous oxide temptations of Bristol, Davy almost died while at the Royal Institution, falling ill in December 1807 and not resuming his scientific work until April 1808. There are various accounts as to what illness Davy had succumbed to, ranging from his falling foul of some of the harmful chemicals that he was working with in the lab, to contracting typhus from London's Newgate prison, where he had been employed to investigate and disinfect the ventilation system. While some scholars have argued that this incident encouraged Davy to redraft 'The Spinosist', publishing it as 'Written after Recovery from a Dangerous Illness', this redrafting (following the notes from Coleridge) had in fact occurred some years earlier, as evidenced by its publication in the December 1806 issue of *The Gentleman's Magazine*.[30] What is still unclear is why Davy chose to publish the poem at this moment in time, and who it was that gave this poem its title, as it is untitled in the original version written in Davy's notebooks. There are only a couple of other instances of Davy publishing poetry during his lifetime, most notably in the prologue that he wrote for *The Honeymoon*, a play by the English playwright John Tobin, in 1805.[31] While he was known to have circulated poetry in manuscript form and to occasionally read it out at parties, the following two extracts from his notebooks (the first written around 1805 and the second around 1815) would seem to suggest that rather than publishing his poetry, he instead chose to achieve an ongoing legacy through his scientific achievements:

The object of poetry, whatever may be said by poets, is more to amuse than instruct; the object of science more to instruct than amuse.

Those brilliant and poetical works in which enthusiasm takes place of reason, and in which the human intellect exhausts itself, as it were, in imagination and feeling, resemble monstrous flowers, brilliant and odorous, but affording no materials of re-production.[32]

In 1812 Davy was knighted for his services to science by the Prince Regent, George IV, and also married the wealthy Scottish widow Jane Apreece, cousin of the Scottish poet and novelist Walter Scott. Another significant event was to occur in this year: Davy's introduction to the young English scientist Michael Faraday. Faraday had attended several of Davy's Royal Institution lectures, taking detailed notes before compiling them into a 300-page essay and sending this to Davy in the hope of a job. Davy wrote back to Faraday stating how impressed he was with his workings, yet without an offer of employment. However, following an accident in the laboratory in 1813, in which Davy nearly blinded himself, he decided that he needed to hire an assistant, and approached Faraday for the role. Faraday's initial task was to serve as Davy's personal valet for a two-year tour of continental Europe. Not only was this a particularly dangerous time for an Englishman to be travelling to Europe (following the French Revolution, and up until about 1815, the two countries were engaged almost continuously in wars), but by all accounts Faraday's life was made even more miserable by the fact that Jane Davy neglected to treat him as an equal, making him eat his food in the servants' quarters and refusing to ride in the same coach as him.

In 1815 Davy returned to England, whereupon he began work on arguably his most famous scientific achievement: the 'Davy lamp'. In those times, working in the mines was an arduous and dangerous occupation, not least because the candles or small lamps that the miners hung from crevices or hammered into timbers near their work could ignite pockets of methane gas that built up over time. In the years leading up to 1815 there had been several such incidents, leading to many fatalities, and causing the Revd Robert Gray of Bishopwearmouth in Sunderland to write to Davy to ask if he might turn his scientific intellect towards coming up with a solution. In many ways this represented the perfect scientific opportunity for Davy, one that with the benefit of hindsight seems almost clichéd. Here was a famous scientist who had grown up in a mining town, who had foreshadowed his own greatness by drawing a picture of a mining lamp wrapped in a wreath on the outside of his first scientific notebook; a Romantic scientist who was striving to demonstrate to a sceptical public the positive ways in which science could benefit the world, and who had self-experimented on the inhalation of toxic gases; a man who himself had almost died, and who had severely hindered his own sight through the unpredictability of chemical substances.

After receiving Gray's request, Davy began experimenting immediately and soon developed a new and fairly simple design; a basic lamp that had a wire mesh built up around it like a chimney, thereby enclosing the flame. The holes of the mesh let the light pass through, but the metal absorbed the heat, meaning that the flame was unable to heat enough methane to cause a dangerous explosion. In November 1815 Davy presented his initial design ideas to the Royal Society, although it wasn't until the following month that he fully developed the idea of a wire mesh. While this research was later published in the *Philosophical Transactions of the Royal Society of London*, he did not patent his invention, ensuring that the lamp could be recreated cheaply, which led to a reduction in the number of mining deaths.[33] For his efforts, Davy was awarded the Rumford Medal by the Royal Society in 1816, a prize awarded every other year for an 'outstandingly important recent discovery in the field of thermal or optical properties of matter made by a scientist working in Europe'.[34]

Despite the success of the Davy lamp, there was controversy surrounding its discovery, with two other inventors claiming to have developed their own versions of a similar safety lamp at around the same time. William Reid Clanny was a Sunderland-based physician who had presented his own version of a safety lamp to the Royal Society in 1813, and while Davy certainly had access to Clanny's lamp, the designs were markedly different. In comparison to the relative simplicity of Davy's lamp, Clanny's involved the use of bellows and water cisterns, and as such was quite cumbersome and not as readily taken up by the coal mining industry.

However, Davy had a much more difficult time convincing people that his lamp had been developed independently of, and before, the work of George Stephenson, the Northumberland-born engineer who would go on to pioneer rail transport. Like Davy, Stephenson had developed a safety lamp that worked by restricting the air flow into the chamber housing the candle, and while the design of the Davy lamp and the 'Geordie lamp' (the moniker attached to Stephenson's invention) had many similarities, they also had differences that might support independent discovery. A fierce debate now raged over who could lay claim to being the true inventor of the safety lamp, with these discussions making it all the way into Parliament, where one of the staunchest defenders of Davy's claim was Lord John Lambton, the First Earl of Durham. In a letter to Lambton, dated 29 October 1816, Davy writes to thank him for his support, making it clear what he thinks of his rivals:

> I never heard a word of George Stevenson[35] & his lamps till six weeks after my principle of security had been published; and the general impression of the scientific men in London, which is confirmed by what I heard at Newcastle was that Stephenson had some loose idea floating in his mind which he had unsuccessfully attempted to put in practice till after my labours were made known; then, he made something like a safe lamp:– but it is <u>not a safe lamp</u>, for the apertures below are four times too large & those above 20 times too large.– But, even if Stevenson's plans had not been posterior to my principles still there is no analogy between his glass exploding machine & my metallic tissue permeable to light and air & impermeable to flame ... Your conduct at no very distant period will be contrasted with that of some other great coal proprietors who find reasons for their indifference as to a benefit conferred upon them in insinuations respecting the claims of Dr. Clanny, Mr. Stevenson etc. When men resolve to be ungrateful it is natural that they should be illiberal & illiberality often hardens into malignity ... I hope you will not blame me for not taking every notice of the attacks of my enemies in the north. I have no desire to go out of my way to crush gnats that buzz at a distance & that do not bite me or to quarrel with persons who shoot arrows at the moon & who believe because they have for an instant intercepted a portion of her light, that they have hit their mark.

From this letter, it is clear that Davy viewed both Clanny and Stephenson with a certain degree of contempt, and despite his protestations he was clearly expending a great deal of effort to crush these 'gnats that buzz at a distance'. Davy seemed to hold special contempt for Stephenson, a self-taught man who had developed his invention through experimentation in the mines and a trial-and-error approach, in contrast to Davy's lab-based work which was rooted in a deep understanding of the chemical processes that were involved; as Davy saw it, Stephenson had no right to be considered in the same breath as one of the true sons of genius. In a letter dated 10 November 1817 to John Buddle, a prominent mining engineer and entrepreneur in the north-east of England who had introduced Davy's safety lamp to all of the collieries under his direction, Davy wrote: 'With regard to Stephenson I have always considered him since I read his letters and pamphlet as an illiterate pirate.'

This controversy continued for several years, and Davy did not find it beneath himself to draw on all the nepotistic support that a man of his standing could muster, calling on the president of the Royal Society (the English naturalist Joseph Banks) to defend his invention, which ultimately led to him defeating Stephenson's claims. Davy's work on the safety lamp brought experimental science to the eyes of the public, and categorically demonstrated how chemistry (and science) could be used to benefit society, thereby justifying Wordsworth's praise of him and the reconsideration of the role of science as a potential force for good. However, Davy himself remained bitter and unhappy with how his role in the development of the lamp had been credited, and despite numerous plaudits he still felt a debt of ingratitude. He harboured these feelings, and later wrote the following poem, which appears in a notebook used between 1819 and 1827, entitled 'Thought after the ingratitude of the Northumbrians with respect to the Safety Lamp', and which reflects many of the sentiments that he expressed in his earlier letter to Lord Lambton:

And though in all my intercourse with man
The feelings recollected scarcely leave

Aught to admire or glory in. Though good
Has been replaced with evil. And a light
Of Science & humanity received
With stern ingratitude — Yet have I not
Recanted, or relaxed in labours high
For these my enemies. And if a chill
Of indignation has oppressed my mind
It was but transitory like the chill
Of a snow cloud in summer. Which though dark
And threatening soon in genial dews
Dissolves to vivify the parched earth.
And I have not both coldness & unkindness
As the cingalian tree[36] which when cut
Does not alone perfume the axe: but gives a balmy oil
Which preserves its harsh & sullen texture from decay. —[37]

Failures and consolations

After the success of his mining lamp, and despite its controversies, Davy was now a well-known and well-respected scientist, and following his earlier trip to mainland Europe, he returned in 1818, on a mission to decipher some ancient Greek philosophical texts. The Herculaneum papyri were found in the eighteenth century in the ruins of Herculaneum, a town in the shadow of Mount Vesuvius that was destroyed when the volcano erupted in AD 79. The papyri were carbonised by the eruption, and while traditional methods of unrolling the scrolls had met with some initial success, opening them would often damage the writing, while exposure to the air could cause the ink to fade. Deciphering the scrolls was a problem that had been tackled by several scholars, and to Davy, a man who considered himself tasked with the responsibility of demonstrating how science could be a force for good, and who from his early childhood had been a keen translator and classicist, this no doubt proved an irresistible challenge. In his laboratory, Davy had some success in using a combination of heat and chlorine to reveal the text on some samples of the papyri that he had obtained, presenting his findings to the Royal Society,[38] and convincing the Prince Regent, George IV, to commission him to travel to Naples to begin work on the Herculaneum papyri themselves.

On his journey to Naples Davy made several stops, including a visit to Rome where he met Lord Byron, whom he had first befriended in 1813 when Davy and his wife had invited the poet to breakfast. Byron was said to have been impressed by Davy, immortalising him in the first Canto of *Don Juan*, verse 132:

> This is the patent-age of new inventions
>   For killing bodies, and for saving souls,
> All propagated with the best intentions;
>   Sir Humphrey[39] Davy's lantern, by which coals
> Are safely mined for in the mode he mentions,
>   Timbuctoo travels, voyages to the Poles,
> Are ways to benefit mankind, as true,
> Perhaps, as shooting them at Waterloo.[40]

Despite Byron crediting Davy as the man most responsible for popularising science and demonstrating to the public what it was capable of, he is less optimistic about what this might entail. Byron was less convinced than Davy, or even Wordsworth, about the positive effect that science might have on society; while science had helped to facilitate the mining of coal, it had also helped to facilitate the killing of men.

For his part, Davy was sufficiently enamoured of Lord Byron to write two poems about him in his notebooks, including 'On the Death of Lord Byron, composed at Westhill in the great storm Nov. 1824',[41] the final verse of which makes it clear that Davy considered Byron to be one of his sons of genius:

> And such may be his fate! And if to bring
> His memory back, an earthly type were given,
> And I possess'd the artist's powerful hand,
> A genius with an eagle's powerful wing
> Should press the earth recumbent, looking on heaven
> With wistful eye; a broken lamp should stand

Beside him, on the ground its naphtha[42] flowing
In the bright flame, o'er earthly ashes glowing.[43]

Upon reaching Naples, Davy was successful in both gaining access to the papyri, through an introduction from the Prince Regent, and in partially unrolling twenty-three of the scrolls, from which some text was also successfully recovered. However, he found the staff at the Naples National Archaeological Museum, where the papyri were held, to be uncooperative and unpleasant, and ultimately drew an end to the experiment. As can be seen from this extract from a letter to the English poet William Sotheby, dated 26 March 1820, he considered the attitude of the Italians, rather than the limitations of his methods, to be the cause of the failures: 'In spite of difficulties, jealousies, & annoyances of every kind, the experiment has been fairly tried: & to have gone on would I think only have been a waste of public money.'

Later that year Davy returned to England, prompted in part by the death of Joseph Banks. Since his appointment in 1778, the Royal Society had been presided over by Banks, who had been widely criticised for a nepotistic approach to awarding fellowships that had resulted in a large cohort of non-scientific members.[44] Following Banks's death on 19 June 1820, the English chemist and physicist William Hyde Wollaston took over as caretaker. Now returned from Italy, Davy was convinced that he was the only man who could unite the warring factions of the Royal Society, and in the November elections of 1820 he was elected to the presidency. However, while he had been elected unopposed, support for Davy was far from unanimous.

One of the first problems that Davy tried to address was the large number of non-scientific members of the Society, many of whom had connections with the British Admiralty and had gained fellowship through Banks's nepotism.[45] These fellows became aggravated when Davy tried to put a cap on the annual fellowship, and so he sought to address this issue by working with his friend John Wilson Croker, then secretary to the Admiralty, to create the Athenaeum Club in 1824, proposing it as an alternative society to which non-scientific men could apply. While the Athenaeum Club was considered a success, there were still many other factions within the Royal Society that Davy had to contend with. On opposite sides stood the Banksians, those supporters who remained loyal to Banks, and a group known as the reformers, those who had vehemently opposed Banks's regime. The Banksians were led by Davies Gilbert,[46] Davy's one-time mentor and a man whom Banks had groomed for the presidency yet who had refused to stand following Banks's death. The reformers were led by Wollaston, a collaborator of Davy's and a man who, despite not standing beyond his role as caretaker, was supported by a group of young and aspiring mathematicians such as Charles Babbage and John Herschel, who doubted that Davy was willing and/or able to push through the reforms that they believed were necessary for the modernisation of science. Davy found himself caught between a rock and a hard place, and to make matters worse he also had to contend with the fellowship of his protégé, Michael Faraday.

Since being hired as Davy's laboratory assistant and personal valet, Faraday had gone on to achieve great success at the Royal Institution, his scientific feats including the discovery of electromagnetic rotation and the exploration of the electromagnetic properties of materials. However, prior to Faraday's successes, Davy and Wollaston had tried, unsuccessfully, to build an electric motor, and Davy thought that Faraday had not properly acknowledged their role in his work. Davy was also aggrieved when Faraday accidentally became the first person to liquefy chlorine, following an experiment that he had conducted at Davy's suggestion. As such, when Faraday was nominated in 1823 to be a Fellow of the Royal Society, Davy opposed his nomination. There was a good political reason for him to deny Faraday's fellowship, as it might have been seen by the reformers as exemplifying the kind of Banksian nepotism that Davy had promised to curtail; undoubtedly, however, this incident further damaged the fragile relationship between the two men.[47]

A further insight into how Davy viewed this relationship might be gleaned from the following poem, written by Davy while on holiday in the Scottish Highlands in 1821, shortly after being elected president of the Royal Society:

> The mighty birds still upward rose,
> In slow but constant and most steady flight,
> The young ones following; and they would pause,
> As if to teach them how to bear the light,
> And keep the solar glory full in sight.
> So went they on till, from excess of pain,
> I could no longer bear the scorching rays;
> And when I looked again they were not seen,
> Lost in the brightness of the solar blaze.
> Their memory left a type, and a desire:
> So should I wish towards the light to rise,
> Instructing younger spirits to aspire
> Where I could never reach amidst the skies,
> And joy below to see them lifted higher,
> Seeking the light of purest glory's prize.
> So would I look on splendour's brightest day
> With an undazzled eye, and steadily
> Soar upwards full in the immortal ray,
> Through the blue depths of the unbounded sky,
> Portraying wisdom's boundless purity.
> Before me still a lingering ray appears,
> But broken and prismatic, seen thro' tears,
> The light of joy and immortality.[48]

In his biography of his brother, John Davy states that this poem was simply about two eagles teaching their young to fly, and that Davy recorded it in verse, as was the case whenever something inspired him. However, as noted by Richard Holmes, this poem can also be read as an allegory for the relationship between Faraday and Davy, beginning with an observation of the older eagles teaching their young how to fly, followed by a reflection from the author (Davy), who initially reveals that that he too wants to inspire greatness in his offspring (Faraday), and that he would find joy in seeing this offspring surpass his own achievements.[49] However, in doing so it is not the immortality of the offspring that would be guaranteed, but that of the parent. The poem reveals further the conflict between Davy the altruist (wanting to see his offspring succeed) and Davy the narcissist (claiming this success as a means to guarantee his own immortality). This duality in Davy's nature is also evident in the remark that he often made when talking about Faraday: that he had been Davy's greatest discovery.

Despite the difficulties that Davy was facing from the Admiralty, the Royal Society still acted as key scientific advisors to the British fleet, and in 1823 it was asked to come up with a scientific solution to stop the copper bottoms of Royal Navy ships from corroding. The normal procedure would have been for Davy to form a committee and develop a collaborative approach to finding a solution, but buoyed by his recent successes, and having had his fingers burned with the safety lamp, he was reluctant to collaborate. Instead he set about trying to devise a solution himself, with Faraday as his assistant. In laboratory experiments, Davy had discovered that pieces of iron or zinc could be used to protect copper through a process known as cathodic protection. The iron or zinc became a 'sacrificial metal', acting as the anode (the electrode from which current flows into an electrical device) in an electrochemical cell with the copper as the cathode (the electrode from which current flows out of an electrical device), with the seawater allowing the flow of current to occur; the result of this was that the iron or zinc became corroded instead of the copper. After proving that this worked in the laboratory, Davy ordered these 'protectors' to be fitted to three ships in Portsmouth dockyard. The initial results proved the protectors to be extremely effective, and with the Admiralty impressed by the tests, and Davy convinced of the further positive benefits that both he and science could offer to society, an order was issued to fit protectors to the entire British fleet.

Unfortunately, what Davy had failed to realise was that the initial tests that he had conducted in the laboratory were not representative of the uncontrolled environment of the open sea, and it soon became apparent that something was wrong, with ships' captains reporting issues with the steering of their vessels. What transpired was that while the protectors worked extremely well in terms of protecting the copper, the electrochemical cell that was created had the side effect of producing nutrients for weeds and barnacles, which made their homes on the bottom of the ships, weighing them down and affecting the steering. By the end of 1825 the Admiralty had stopped production of the protectors, and had ordered all the protected ships to have their sacrificial metals removed. This was an embarrassing and very costly failure for both the Admiralty and Davy, whose decision to 'go it alone' had backfired spectacularly.[50]

At this point in his life, Davy found himself beset by two types of disappointment: the inability to convince political allies and the failure to generate scientific answers. Both failures were alien to him, and no doubt contributed to his suffering a stroke in 1827, following which he departed for Italy to improve his health, giving up his presidency of the Royal Society in the process.[51]

Davy spent the last two years of his life travelling around mainland Europe, visiting friends, and writing *Salmonia*, a book on fishing, and *Consolations in Travel*, a mishmash of poetry, scientific observation, and philosophical musings, which was published posthumously.[52] In *Consolations in Travel*, Davy begins by taking us on a tour of our solar system, albeit one that is populated by fantastical creatures and led by a supernatural 'Genius'. There are then some discussions of various Italian historical and natural landmarks, and it concludes with some more open-ended observations about the mysteries of nature and the role that science can play in positively benefiting humankind. While *Consolations in Travel* lacks a central narrative, as a collection of dialogues it was very well received, influencing, among others, Anne Brontë, who references the work in *The Tenant of Wildfell Hall*.[53]

Davy was now probably writing to ensure his legacy, while also hoping to positively benefit society as he had sought to do with his scientific achievements. This willingness (and need) to serve *his* public is reflected in the preface to *Consolations in Travel*, written by Davy himself, which ends with:

He [Davy] has derived some pleasure and some consolation, when most other sources of consolation and pleasure were closed to him, from this exercise of his mind; and, he ventures to hope that these hours of sickness may be not altogether unprofitable to persons in perfect health.[54]

As his body degenerated, Davy believed that his mind was becoming sharper, and in coming to terms with his sickness and the inevitability of death, he revisited some of the successes and failings of his earlier career; there is an especially poignant section in the fifth dialogue of the book, 'The Chemical Philosopher', in which Davy argues for the importance of a methodical approach to scientific investigation:

> By often repeating a process or an observation, the errors connected with hasty operations or imperfect views are annihilated; and, provided the assistant has no preconceived notions of his own, and is ignorant of the object of his employer in making the experiment, his simple and bare detail of facts will often be the best foundation for an opinion.[55]

It is difficult to read this passage without being reminded of the 'hasty operations' that Davy employed when developing the protectors for the Admiralty, nor to detect his grumbling admiration of Faraday, providing of course that *his* assistant knew his place, and was aware of the role that Davy had played in his development. The trial-and-error approach that was favoured by Faraday is the same methodology for which Davy originally castigated Stephenson, and is perhaps another reason for the begrudging attitude that Davy felt towards his erstwhile apprentice.

During his final trip across the continent, Davy continued to make use of his notebooks. However, while earlier notebooks were quite similar in their structure to *Consolations in Travel* (containing a mix of poetry, scientific observations, and philosophical thoughts), Davy used his final notebook to record an almost continuous series of poems, both originals and redrafts of his earlier work. Many of these poems return to and develop some of the ideas that he had originally ruminated on while in Bristol, particularly the role of nature and the potential for immortality that it offered even dying men. These thoughts are perhaps best exemplified in the following poem, written in 1825 without a title in Davy's final notebook:

And when the light of life is flying
And darkness round us seems to close
Nought we truly know of dying
Save sinking in a deep repose

And as in sweetest soundest slumber
The mind enjoys it highest dreams
And as in stillest night we number
Thousands of worlds in starlight beams

So may we hope the undying spirit
In quitting its decaying form
Breaks forth new glory to inherit
As lightning from the gloomy storm.[56]

An influential end

Davy died on Friday 29 May 1829, in a hotel room in Geneva, Switzerland. He was laid to rest in the Plainpalais cemetery (also known as the Cemetery of Kings) just outside Geneva. As well as having requested to be buried where he died, Davy was also concerned that he might be comatose rather than fully dead, and so had asked for his burial to be delayed and for an autopsy to be dispensed with. However, the laws of Geneva forbade such a delay, and so he was buried on the Monday following his death, an incident that Davy would no doubt have viewed as a personal slight.

Since his death, Davy has gone on to influence several literary works, not least *Frankenstein; or, The Modern Prometheus* by Mary Shelley, who was known to be a keen admirer of Davy, with some of the statements by Professor Waldman, Victor Frankenstein's professor of chemistry at the University of Ingolstadt, being almost direct quotations from Davy's works.[57] Davy was also further immortalised in print by Edmund Clerihew Bentley, the

English novelist and humourist who invented the clerihew – a brief biography in the style of a humorous verse, typically in two rhyming couplets with lines of unequal length – and who chose Davy as the topic of his first published poem in this style:

Sir Humphry Davy
Detested gravy.
He lived in the odium
Of having discovered sodium.[58]

As with Coryton's poem that appears at the beginning of this chapter, Bentley toys with the assonance of Davy's surname to poke fun at his subject. Bentley is said to have written the first two lines of this poem while in a chemistry class at school, perhaps at the same time as learning about Davy's scientific achievements, including the first isolation of sodium. While Bentley devised the clerihew as an exercise in finding silly rhymes for famous people's names, the last two lines of the poem point to a truth that Davy grapples with in some of his later writings – that even given his great accomplishments in science, were they enough?

Despite his pettiness, narcissism and lack of tact, Davy had, through his scientific achievements, presented a new future vision of humanity, one which benefited from, rather than was torn apart by, science. Today, Davy is remembered as one of the founders of modern chemistry, and as a poet who influenced some of the thinking of the Romantic movement, not just through his poetry but through his pursuit of science as a force for public good; something that could ultimately help benefit nature and in turn humankind. And whether through his youthful follies with nitrous oxide, his successes with electrochemistry, or his contemplation of nature, Davy was throughout his life equal parts lyre and lamp; a true Son of Genius who in death found the immortality that he had so desperately craved in life.

2

The metaphysical poet: Ada Lovelace

A hidden light may burn that never dies,
But bursts thro storms in purest hues exprest.
From 'The Rainbow' by Ada Lovelace[1]

Born in bitterness and nurtured in convulsion

Augusta Ada Byron was born on 10 December 1815 to Anne Isabella (Annabella) Byron and George Gordon, Lord Byron, the celebrated Romantic poet.[2] Her parents did not have what could be considered an easy relationship, beginning with a troubled court-ship that resulted in a failed proposal and caused Byron to note of Annabella that 'Her proceedings are quite rectangular, or rather we are two parallel lines prolonged to infinity side by side never to meet.'[3] Despite these reservations, and probably because of his growing debts (Annabella came from a wealthy and prominent family), Byron proposed again and the couple were married on 2 January 1815. However, from the onset, Annabella was convinced that her husband was committing incest with his half-sister Augusta Leigh (as well as carrying on numerous other affairs), which as a God-fearing woman was something that troubled her greatly. It was clearly not a happy atmosphere into which Lovelace was born, causing her father to later write that she was: 'The child of love, – though born in bitterness, and nurtured in convulsion.'[4] Alongside her suspicions about Byron's incest and adultery, Annabella also considered her husband to be utterly insane and a danger to her young daughter, so only a few weeks after giving birth she moved to Kirkby Hall in Leicester to stay with her parents, taking her daughter with her and instructing her father Ralph Milbanke to ask Byron for a separation.

It was very difficult for women of that era to force through a separation, and it was also unusual for mothers to gain sole custody of their children. However, through a series of thinly veiled threats to Byron, Annabella made it perfectly clear that she would be willing to make public the allegations of incest. These threats, the separation, and his still mounting debts caused Byron to leave England forever in April 1816. He would die just eight years later in Missolonghi, from a fever contracted while fighting in the Greek War of Independence, having never again seen his daughter.

Annabella was adamant that her daughter would not turn into the 'mad, bad, and dangerous to know' poet[5] that her father had become, and set about devising a scheme of private tuition that was focused mainly around mathematics and science. Such an education was very unusual for a young lady of that era, although Annabella herself had been a gifted and only child whose indulgent parents had arranged for her to be tutored by the ex-Cambridge tutor William Frend. Annabella's intelligence and love of maths was well known and later mocked by her estranged husband. When they first began courting he referred to her as an 'amiable mathematician', which later became the 'Princess of Parallelograms' and finally the 'Mathematical Medea' as their relationship soured. After their separation, Byron further mocked Annabella's arithmetical intellect through his poetry, portraying her as the protagonist's mother Donna Inez in his satirical poem *Don Juan*, where she is first introduced near the beginning of the poem as a pious, smart and frigid woman:

> Her favourite science was the mathematical,
>   Her noblest virtue was her magnanimity,
> Her wit (she sometimes tried at wit) was Attic all,
>   Her serious sayings darken'd to sublimity;
> In short, in all things she was fairly what I call
>   A prodigy – her morning dress was dimity

Her evening silk, or, in the summer, muslin,
   And other stuffs, with which I won't stay puzzling.[6]

Further steps taken by Annabella to banish the Byron temperament included the insistence that her daughter spend prolonged periods of time lying perfectly still, to improve her self-control. Throughout her childhood tuition, Lovelace proved herself to be an extremely capable pupil, and her high social status meant that she was regularly in contact with the various scientific and literary luminaries of the day, from Charles Dickens and Michael Faraday to Augustus De Morgan and Mary Somerville, with De Morgan and Somerville (a noted mathematician and astronomer, respectively) later becoming Lovelace's tutors at various points during her life. However, no other figure of that era would have a greater influence on her life than the English mathematician and engineer Charles Babbage.

Analysis and metaphysics

When Lovelace first met Babbage in June 1833 (she was 17 at the time) he had been working on his Difference Engine for over a decade. The purpose of this engine was to automatically compute values of polynomial functions, and it represented the first complete design for an automatic calculating engine. While the engine was yet to be finished (indeed it was never completed in Babbage's lifetime, and was only fully constructed for the first time in the late twentieth century), a small section of it was in working order and could be used to demonstrate its principles. Shortly after the two met for the first time, Babbage invited Lovelace to see this section of the Difference Engine in action. Unfortunately, around this time the government funding that he had received to develop the engine was running out, and while Babbage tried to rein in the costs, this ultimately led to a falling out with his chief engineer, Joseph Clement, who left and took all the designs with him as severance.

Babbage was not to be deterred and had already begun working on a new machine: the Analytical Engine. Unlike the Difference

Engine, which was only capable of calculating specific polynomial functions, the Analytical Engine would be capable of many different computational tasks. The designs of this Analytical Engine (which was also never fully built by Babbage) share many of the common properties of modern computers, including hardcopy printouts, graph plotting capabilities and the potential for variables to be inputted via punched cards.

Babbage was unsuccessful in leveraging more funds from the British government to build his Analytical Engine, which is perhaps unsurprising given the fact that he was yet to build the Difference Engine for which the government *had* provided funding. Undeterred, and encouraged by Lovelace (with whom he was now in regular correspondence), Babbage visited Italy in 1840, where he was invited to speak to a group of engineers in Turin about his new machine. Among the audience at his lectures was the 30-year-old Luigi Menabrea, an officer of the military engineers who would later become the prime minister of Italy. Menabrea used his notes to prepare a scientific paper for publication about the principles of the Analytical Engine, which two years after Babbage's trip to Turin was written up in French and published in the Swiss journal *Bibliothèque universelle de Genève* in 1842. Up until this point there had been no other publications concerning the Analytical Machine, and so Babbage's friend, the English scientist and inventor Charles Wheatstone, commissioned Lovelace to translate Menabrea's paper into English.

During the translation, Lovelace produced a series of notes to go alongside Menabrea's work, as she felt that there were certain things that needed to be explained in more detail, and others that required further analysis. The notes that she provided eventually ran to three times the original length of Menabrea's writings, and were simply signed 'A.A.L.'[7] Arguably the most important contribution can be found in Note G, where Lovelace sets out the concept of designing a program that can be inputted into the Analytical Engine to calculate a series of Bernoulli numbers. Bernoulli numbers are essentially a sequence of numbers that are used in the evaluation of a variety of functions, and that Lovelace laid out the process by which the Analytical Engine could calculate them is important for two reasons. First, it is the first explicit example of an algorithm that was specifically tailored to be run by a computer, thereby making it the first computer program and Lovelace the first computer programmer. Secondly, it demonstrated that she conceptually understood the universal computing power of Babbage's machine, that is, the notion that despite having a fixed set of hardware, by simply altering the software it could perform different tasks.

In recent years there have been several attempts to detract from Lovelace's accomplishments. These disagreements essentially boil down to two arguments: that Lovelace was not intelligent enough to have developed such a program without significant help from Babbage, and that the Bernoulli program was pre-dated by several earlier programs written by Babbage in his notebook. However, from the letters between Lovelace and Babbage (a selection of which are expertly transcribed and contextualised in Betty Toole's *Ada, the Enchantress of Numbers*) it is clear that this is not the case.[8] The fact that Lovelace was the first person to come up with the idea of using Bernoulli's numbers to demonstrate the programming capabilities of the Analytical Engine is evident from this letter that she sent to Babbage on 10 July 1843: 'I want to put in something about Bernoulli's Numbers, in one of my Notes, as an example of how an implicit function, may be worked out by the engine, without having been worked out by human head & hands first.'[9]

From the ensuing correspondence, it is apparent that Lovelace is beta-testing her ideas with Babbage, in some instances asking him to provide technical checks, not unlike the process by which a modern software engineer would develop a new program for a client's specific set of hardware; that is, Lovelace is the programmer and Babbage the client. With regard to whether Babbage's own notes pre-date the Bernoulli program proposed by Lovelace, it is true that they do contain the concepts of some basic programs, but none of them are as intricate or as fully formed as Lovelace's. There is further evidence from Babbage's autobiography, *Passages from the Life of a Philosopher*, that the introduction of the Bernoulli program (which Babbage refers to as an illustration) was an original idea from Lovelace: 'We discussed together the various illustrations that might be introduced: I suggested several, but the selection was entirely her own.'[10] Here, Babbage is not implying that he simply laid out a selection of suitable ideas for Lovelace to select; rather, he is stating that despite offering a few suggestions for topics that might work, Lovelace decided on the course of development for the Bernoulli program entirely of her own volition.

In addition to the introduction of the Bernoulli program, the notes that Lovelace provided to the Menabrea translation offer a wealth of metaphysical and analytical insight, with two of the most remarkable pieces of commentary appearing in Notes A and G, respectively:

> Supposing, for instance, that the fundamental relations of pitched sounds in the science of harmony and of musical composition were susceptible of such expression and adaptations, the engine might compose elaborate and scientific pieces of music of any degree of complexity or extent.

> The Analytical Engine has no pretensions whatever to originate anything. It can do whatever we know how to order it to perform. It can follow analysis; but it has no power of anticipating any analytical relations or truths.[11]

With regards to the observation that appeared in Note G, Lovelace was effectively acknowledging the limitations of the Analytical Engine, and calling into question the concept of Artificial Intelligence; something that the English mathematician Alan Turing would later term 'Lady Lovelace's Objection' in his seminal work on the subject.[12]

The observation that appears in Note A is extraordinary. In it, Lovelace explores the notion that, were the Analytical Engine to be inputted with musical sounds and structures, it would be capable of composing music according to the rules with which it was programmed. This is an astonishing insight, essentially foreshadowing the formalisation of the universal computer, which would not be done by Turing for almost another century,[13] and further demonstrating that it was Lovelace, and not Babbage, who fully grasped the metaphysical implications of what the Analytical Engine was truly capable of.

As well as the direct translation from French into English, the accompanying notes provided a translation of the mechanical notation used by Babbage into a language that could be read and understood by others. However, it was a misunderstanding of what this role of interpreter entailed which led to a falling out between Lovelace and Babbage prior to the initial publication of the translation and notes. Initially, Babbage wanted Lovelace to include an unsigned preface, written by himself, which effectively outlined the stupidity and lack of foresight of the British government in not providing the necessary funds to enable the Analytical Engine to be built. Lovelace opposed this, as she believed that to publish such a preface would be political suicide; she wanted to help the Analytical Engine become a reality, but she also believed that Babbage was a 'poor player at the political game'.[14] Babbage then demanded that Lovelace retract the manuscript, which she did not do, causing him to write to her in a letter dated 8 August 1843: 'Had the Editor been in England I believe he would at my request have inserted my defence and forborn to have printed the paper.'[15]

Thankfully, the feud was short-lived, and the manuscript went on to be published. Just over a month later, on 9 September 1843, Babbage wrote to Lovelace expressing his desire to spend time in her company at Ashley (a Lovelace family property in Somerset), so that he could 'Forget this world and all its troubles and if possible its multitudinous charlatans – everything in short but the Enchantress of Number.'[16] As well as referring to her as the 'Enchantress of Number', he also signed off this letter with: 'Farewell my dear and much admired interpreter'. Babbage's explicit acknowledgement of Lovelace as his interpreter was pronounced again on 12 September 1843, when he signed his letter: 'Ever my fair interpretress'.[17] Shortly after this, in a letter to her mother dated 15 September 1853, Lovelace herself acknowledged her role as the interpreter of Babbage's work, declaring herself to be the 'High-Priestess of Babbage's Engine'.[18]

Babbage saw in Lovelace not only a person of intellectual brilliance, but also someone who possessed what he lacked: a way to communicate complex ideas in an engaging and informative manner. In addition to his respect for Lovelace as a scientist and a communicator, Babbage was obviously aware of her father (as *everyone* would have been at the time), and was sensitive to the power of poetry as a medium for effective communication, as evidenced by the following tongue-in-cheek letter that he sent to the English poet Alfred Tennyson, after reading his poem 'The Vision of Sin':

> In your otherwise beautiful poem there is a verse which reads:
>
> > Every moment dies a man,
> > Every moment one is born.
>
> It must be manifest that, were this true, the population of the world would be at a standstill. In truth, the rate of birth is slightly in excess of that of death. I would suggest that in the next edition of your poem you have it read:
>
> > Every moment dies a man
> > Every moment $1\frac{1}{16}$ is born
>
> Strictly speaking this is not correct. The actual figure is a decimal so long that I cannot get it in the line, but I believe $1\frac{1}{16}$ will be sufficiently accurate for poetry. I am, etc.[19]

In reading this, cynics might think that Babbage used Lovelace to interpret his work into a language that others could understand, and used her position in society to further advance his own political and social standing. However, from reading the correspondence between them, it is clear that they were close friends who had a deep respect and fondness for one another.[20] Babbage himself never published anything about the Analytical Engine until his autobiography in 1864, where a chapter on the subject appeared prefaced by the following quote from 'The Prophecy of Dante', a poem that was written by Lovelace's father: 'Man wrongs, and Time avenges'.[21]

Like father, like daughter

Because of his self-imposed exile and early demise, Lord Byron played no direct part in his daughter's upbringing. However, it is clear from his poetry and writings that she weighed heavily upon his mind, as evidenced by the opening stanza from the third Canto of his well-loved *Childe Harold's Pilgrimage*, first published in 1816, shortly after Byron had left England for good:

Is thy face like thy mother's, my fair child!
Ada! sole daughter of my house and heart?
When last I saw thy young blue eyes they smiled,
And then we parted, – not as now we part,
But with a hope. –
                    Awaking with a start,
The waters heave around me; and on high
The winds lift up their voices: I depart,
Whither I know not; but the hour's gone by,
When Albion's lessening shores could grieve or glad mine eye.[22]

Byron makes it clear that his daughter is to be his muse, his inspiration in describing the battles and the heroic feats that make up the subject matter in the remainder of this Canto. And it is in the final stanza that Byron makes the remark that was quoted at the beginning of this chapter, ending this section of the poem with a blessing for his daughter, and a wish for what might have been:

The child of love, – though born in bitterness,
And nurtured in convulsion, – of thy sire
These were the elements, – and thine no less.
As yet such are around thee, – but thy fire
Shall be more tempered, and thy hope far higher.

Sweet be thy cradled slumbers! O'er the sea,
And from the mountains where I now respire,
Fain would I waft such blessing upon thee,
As, with a sigh, I deem thou might'st have been to me![23]

Despite the absence of her father, and the express wishes of her mother, Lovelace spent various stages of her life exploring her identity as either a mathematician or a poet, often remarking on and exploring the intersections where the two disciplines met and overlapped, as demonstrated in this extract from a letter she wrote to Annabella, dated 11 January 1841:

> I have had a *mathematical*[24] week since I last wrote … and I have made some curious observations as to the *effects* of the study. The principles are as follows: immense development of *imagination*; so much so, that I feel no doubt if I continue my studies I shall in due time be a *Poet*.[25]

She further explored this notion of imagination in an essay that she wrote on 5 January 1841, part of which reads:

> *What* is imagination? We talk *much* of Imagination. We talk of the Imagination of Poets, the Imagination of Artists etc.; I am inclined to think that in general we don't know very exactly *what* we are talking about…
>
> Those who have learned to walk on the threshold of the unknown worlds, by means of what are commonly termed par excellence the exact sciences,[26] may then with the fair white wings of Imagination hope to soar further into the unexplored amidst which we live.[27]

In this essay, Lovelace is arguing that imagination is not the sole preserve of artists or poets, but rather that true imagination only arises when conceptualising the world using the principles of science and mathematics. This thought was echoed by the American physicist Richard Feynman over a hundred years later when he wrote: 'What men are poets who can speak of Jupiter if he were like a man, but if he is an immense spinning sphere of methane and ammonia must be silent?'[28]

Despite her protestations about the superiority of the 'exact sciences', Lovelace still considered herself to be a capable writer, something that she declared to her husband William (in 1835 Ada married William King, who was made Earl of Lovelace in 1838, with Ada becoming Countess of Lovelace) in a letter dated April 1842:

> That I *can* write, I know because I *have* written. And I write too with great facility ... Just to show you that I *can* write, I send you a little thing which I did the other day. It was suggested to me by a German ballad of Schiller's, but I cannot call it either a *translation*; for (except the first stanza, which is rather close), it is *entirely different* from the original.[29]

Here, Lovelace's interpretation of Schiller's ballad[30] has close parallels with her later work on Menabrea's document – something which could not rightly be considered *only* a translation either. Despite Lovelace's writings, both William and Annabella disapproved of her desire to pursue her poetic inclinations, and often strongly encouraged her to return to pursuits of a more scientific persuasion. Yet despite her family's protestations, it was Lovelace's ability and willingness to explore the subjectivism of poetry within the objectivism of her scientific and mathematical education that enabled her to understand the metaphysical potential of Babbage's Analytical Engine. In a letter to Babbage dated 30 July 1843 she notes that: 'I do *not* believe that my father was (or ever could have been) such a *Poet* as *I shall* be an *Analyst*; (& Metaphysician); for with me the two go together indissolubly.'[31]

Despite these protestations, Lovelace respected her father's poetic talent greatly, and just as she was the muse for Canto III of *Childe Harold's Pilgrimage*, the poetic sensibilities of her father probably inspired several of the insights into universal computing that she foresaw. This gender reversal, in which a male figure becomes the muse to a female protagonist, would no doubt have pleased Byron, whose own *Don Juan* inverted traditional gender roles by featuring a passive male lead who then finds himself the target of sexual advances from several female characters.

As part of her poetic output, Lovelace composed odes to family members and friends, one of whom was the English social reformer and founder of modern nursing, Florence Nightingale. The two friends had moved in similar social circles since their youth (Lovelace was six years older) due to the good friendship of their mothers, and in 1847 Nightingale (who at the time would have been around 27) and her father were invited to stay with the Lovelaces. During this time Lovelace distributed a poem that she had written to commemorate the brilliance of her friend, the first stanza of which reads:

> I saw her pass, and paused to think!
> She moves as one on whom to gaze
> With calm and holy thoughts, that link
> The soul to God in prayer and praise.
> She walks as if on heaven's brink,
> Unscathed thro' life's entangled maze.[32]

The reverence that Lovelace shows to her friend is reminiscent in parts of one of Byron's most celebrated works, 'She Walks in Beauty':

> She walks in beauty, like the night
>   Of cloudless climes and starry skies;
> And all that's best of dark and bright
>   Meet in her aspect and her eyes:
> Thus mellow'd to that tender light
>   Which heaven to gaudy day denies.
>
> One shade the more, one ray the less,
>   Had half impaired the nameless grace
> Which waves in every raven tress,
>   Or softly lightens o'er her face;
> Where thoughts serenely sweet express
>   How pure, how dear their dwelling place.
>
> And on that cheek, and o'er that brow,
>   So soft, so calm, yet eloquent,

The smiles that win, the tints that glow,
  But tell of days in goodness spent,
A mind at peace with all below,
  A heart whose love is innocent![33]

This poem is about the extraordinary beauty of a young woman and was written by Byron in response to seeing Anne Beatrix Wilmot (the wife of his first cousin, Robert Wilmot) at a party that he attended. It is a poem that celebrates female beauty, and while Byron is at pains to point out that this also includes the innate 'goodness' of her inner beauty, this feature is only inferred from the woman's smiles and blushes rather than by any acts or accomplishments. To him the woman in the poem is an object to be admired.

Contrast this to the third stanza in Lovelace's poem about Nightingale, where she writes:

And books she loves, and wisdom's lore;
For there her thoughtful nature feels
The priceless treasure held in store,
Which to her earnest mind reveals
Those deeper truths that few explore,
And busy life too oft conceals.[34]

Nightingale was an accomplished statistician, whose 1858 book *Notes on Matters Affecting the Health, Efficiency, and Hospital Administration of the British Army* provided statistical evidence that helped to make the link between mortality rates and hospital conditions. Lovelace was keenly aware of her friend's great intellect, as well as her compassion, and saw in her a kindred spirit; a female from a privileged upbringing who had chosen to pursue scientific and mathematical interests out of a desire to better understand the world in which she lived. In her poem, Lovelace celebrated Nightingale as a person to be respected and loved, and not simply an object to be admired and coveted. While Byron's poem is perhaps the more aesthetically pleasing of the two, it is Lovelace's poem that presents a more well-rounded and analytical representation of the subject. The woman who features in 'She Walks in

Beauty' could be anyone and no one, such is the monument that Byron erects to this fictitious Venus. On the contrary, from her final prophetic stanza (written several years before Nightingale's work in the Crimean War), it is clear to anyone reading Lovelace's poem that she could only have been writing about one woman – Florence Nightingale:

> In future years, in distant climes,
> Should war's dread strife its victims claim,
> Should pestilence, unchecked betimes,
> Strike more than sword, than cannon maim,
> He who then reads these truthful rhymes
> Will trace her progress to undying fame.[35]

This difference between Lovelace the analyst and Byron the poet was further exemplified by their attitudes towards technology, particularly machinery and automation. Despite his familiarity and respect for scientists such as Humphry Davy (see Chapter 1) and his own personal interest in astronomy, like many of the Romantic poets Byron was concerned with the role that technology would play in shaping the future of humankind. In 1812, and just two weeks before he found national fame through the publication of *Childe Harold's Pilgrimage*, Byron used his position as the inheritor of a peerage[36] to make his maiden speech in the House of Lords, in which he outlined that he was strongly against the recently passed Frame Breaking Act. This act decreed that 'frame breakers' should receive the death penalty for breaking the textile machines that were putting them out of work, and Byron was one of the few people in Parliament to stand up in defence of these workers, declaring that:

> Considerable injury has been done to the proprietors of the improved frames. These machines were to them an advantage, inasmuch as they superseded the necessity of employing a number of workmen, who were left in consequence to starve.[37]

The Jacquard looms that were at the centre of this controversy used punched cards to determine the patterns to be woven in the textiles, an idea that later inspired Babbage to utilise a similar mechanism in his Analytical Engine, the physical process through which the Bernoulli program that Lovelace developed could theoretically be implemented by the machine. However, as other scholars have pointed out, the Jacquard loom itself was not the ancestor of today's universal computer.[38] Clearly a loom can only ever process and manufacture textiles, no matter what is punched into the card. In contrast to this, the Analytical Engine that Babbage had designed could perform many different tasks without changing the underlying hardware, thus making it the first example of a machine theoretically capable of universal computation. Nevertheless, Babbage had not designed the Analytical Engine to be a universal computer; he had set out to create a machine that could produce a series of mathematical tables in the most effective way possible. While Babbage had designed the Analytical Engine, it was Lovelace who first realised its potential to go beyond simply calculating numbers. Similarly, while Byron saw the automation of the looms as a symbol of oppression, Lovelace saw the automation of the Analytical Engine as a creative tool that had the potential not only to calculate numbers, but also to create music.

Both Lovelace and her father died when they were 36, their deaths both caused by illnesses that were probably exacerbated by the blood-letting performed by their physicians. As he lay dying from fever in 1824, Byron made a final plea for his daughter's forgiveness and health, remarking to his valet Fletcher: 'Oh, my poor dear child! – my dear Ada! My God! Could I but have seen her! Give her my blessing.'[39] In her final days, Lovelace's mind was similarly on her father; one of her dying wishes (told to her husband, William) was to be laid to rest next to her father in the Byron family tomb.

'The Rainbow'

When Lovelace died, she was indeed laid to rest beside her father in the Church of St Mary Magdalene in Hucknall, Nottinghamshire. Neither Annabella nor Babbage attended the funeral, although in Babbage's case this was probably because, at Annabella's insistence, he, like many of her other friends and contemporaries, was barred from seeing Lovelace towards the end of her life. Lovelace had also requested that her tomb be inscribed with a poem that she had written, entitled 'The Rainbow'. This request was not fulfilled, and a fairly standard tablet marked her grave; the only remarkable thing about her burial chamber was that the lid of her coffin was inscribed with the Lovelace family motto: *Labor ipse voluptas*, which translated from Latin reads: 'labour is its own reward'.

In addition to the tomb in the Church of St Mary Magdalene, Annabella also paid to have a monument built to her daughter in All Saints Church in Kirkby Mallory, Leicestershire, the ancestral resting place of Annabella's family, and close to Kirkby Hall, where Lovelace had spent her childhood. This monument does display a version of 'The Rainbow', my transcription of which is taken from Lovelace's original draft and notes:

> Bow down in hope, in thanks, all ye who mourn:
> Whene'er that peerless arch of radiant hues,
> Surpassing earthly tints, the storm subdues:
> Of nature's strife and tears, 'tis heaven-born,
> To soothe the sad, the sinning, and forlorn;
> A lovely, loving token, to infuse
> The hope, the faith, that Power divine endues
> With latent good the woes by which we're torn.
> 'Tis like a sweet repentance of the skies,
> To beckon all by sense of sin opprest.
> Revealing harmony from tears and sighs;
> A pledge that deep implanted in the breast,
> A hidden light may burn that never dies,
> But bursts thro storms in purest hues exprest.[40]

'The Rainbow' was probably written some time between 1851 and 1852, when Lovelace was suffering from the illness (now thought to be uterine cancer) that would eventually kill her. For most of this time she was bedridden, and prescribed laudanum and other opiates to help with her pain, which at times (not unsurprisingly) greatly affected her lucidity. Some of the imagery and phrases that appear in her letters around this time support this claim for the poem's chronology and suggest that it was probably written around the summer and autumn of 1851. For example, in a letter to her mother, dated 10 August 1851, she states that: 'The gathering of the "*drops*" proceeds, & bright & beautiful drops they appear to me to be; – not *tear* drops, but sparkling with life and light!'[41] In another letter to her mother dated 16 August 1851 she continues to speak about the 'drops', noting that: 'They are *prismatic* drops full of bright & various hues.'[42]

The religious allegories in 'The Rainbow' would probably have appeased her mother, while the use of an Italian poetical structure (the poem is an Italian or Petrarchan sonnet) would no doubt have appealed to her father, who travelled extensively through Italy and wrote in letters to Annabella that he was keen for his daughter to learn the language. However, there is perhaps a more deliberate reason why Lovelace chose to write about a rainbow, and why she was so insistent that this poem should appear on her grave-stone. This was because the rainbow represented a great scientific puzzle that had occupied her mind as a younger student, and that had marked a turning point in her education; the point at which learning about science and maths no longer felt like a command from her mother, but had instead become something that she took great personal pleasure from. In a letter dated 15 March 1834 (when Lovelace would have been 18), she wrote to her mother's old tutor, William Frend:

> I shall be very grateful if you will be kind enough the first time you have a few spare moments, to write me a letter about rainbows. I am very much interested on the subject just now, but I cannot make out one thing at all, viz: why a rainbow always appears to the spectator

to be an arc of a circle. Why is it a curve at all, and why a circle rather than any other curve? I believe I clearly understand *how* it is that the colours are separated, and the different angles which the different colours must make with the original incident ray. I am not sure that I entirely understand the *secondary* rainbow.[43]

When she wrote this letter, Lovelace's tutor was not Frend, but the English physician William King, whom Annabella had instructed to act as her daughter's moral guide as well as her maths tutor. However, King was not always able to provide the mathematical insight that Lovelace required, and so she stopped asking him questions of this nature and instead relied upon her own reasoning, and where necessary wrote to more trusted sources, such as Frend. The rainbow was therefore not only a symbol of defiance against her final illness, it also symbolised the beginning of her scientific interest and independent intelligence. It was this independent intelligence that gave Lovelace the confidence to converse with Babbage in the first instance, and allowed her to develop the notes that accompanied the Menabrea translation.

The positive and life-affirming message that Lovelace presents to us in 'The Rainbow' is worth contrasting with one of her father's most famous poems, 'Darkness', which begins:

I had a dream, which was not all a dream.
The bright sun was extinguish'd, and the stars
Did wander darkling in the eternal space,
Rayless, and pathless, and the icy earth
Swung blind and blackening in the moonless air;
Morn came, and went – and came, and brought no day,
And men forgot their passions in the dread
Of this their desolation; and all hearts
Were chill'd into a selfish prayer for light:[44]

Byron wrote this in July 1816, during the 'Year without a Summer', so called because of the severe climate irregularities and drop in global temperatures that occurred that year. These climatic changes were caused by the ash released into the atmosphere during the

eruption of Mount Tambora in Indonesia in April 1815, the largest
ash eruption since the Ice Age, which resulted in failed harvests and
famine across North America, Asia, and Europe.[45] Byron claimed
that it was this event that inspired him to write 'Darkness', but it
can also be viewed as an apocalyptic poem with many religious
symbols contained within it. Like his daughter's sonnet, Byron had
written a poem that was inspired by both scientific observation and
religious allegory. However, while 'The Rainbow' urges people
to rejoice in the covenant that was made with God after the Flood,
'Darkness' presents an alternative apocalyptic event in which
darkness and famine result in the death not only of everyone on
Earth, but of the universe itself. Byron presents a world in which
all efforts are futile, whereas Lovelace presents one in which there
is always hope.

Lovelace made several different versions and edits of 'The
Rainbow'. However, none of them are an exact match for the version that is given in this chapter, which should be considered to be
her most complete version of the poem, as supported by an analysis
of an earlier draft in the British Library archives.[46] On Lovelace's
monument in All Saints Church, above the poem, there is also the
following inscription:

Inscribed
by the express direction of
Ada Augusta Lovelace
Born Dec 10th 1816: Died Nov 27th 1852
to recall her Memory.

'And the prayer of faith shall save the sick, & the Lord shall
raise him up and if he have committed sins,
they shall be forgiven him'
James V. 15.

The first thing to note about this inscription is that it incorrectly
gives the year of Lovelace's birth as 1816 rather than 1815.
Secondly, there is no record in any of her correspondence that
she asked for this Bible verse to be inscribed upon her tombstone,

and while she was religious, the insertion of such a verse seems more like the work of Annabella. As Lovelace lay dying it was her mother who insisted on being the lone presence at her bed-side, believing that her daughter's suffering and pain represented a pathway to paradise, and the verses that appear in the Bible immediately preceding those quoted also talk about the need for 'patience in suffering'. The appearance of such a passage on some-one's monument would seem to suggest that the deceased had committed sins for which they required absolution. During her life, and in the years since her passing, there have been persistent rumours that Lovelace both conducted numerous affairs and ran up large debts because of a gambling addiction. However, these rumours have been greatly exaggerated, are not based on any concrete primary sources, and were to a large extent embroidered by Annabella. In her final days, Lovelace was coerced by her mother to reveal her 'sins' to William, which resulted in him dis-tancing himself from his wife and handing over all of Lovelace's responsibilities to Annabella. The inclusion of this biblical pas-sage might be interpreted as a further besmirching of Lovelace by her mother, acting to detract from the legacy that is inherent within 'The Rainbow'. This additional 'note' is in stark contrast to the improvements made to the original Menabrea article by Lovelace's own notes and translations.

In a letter to her mother, dated 15 October 1851, Lovelace wrote: 'But I do dread that horrible *struggle*, which I fear is the Byron blood.'[47] As well as confiding in her mother her fears about her own mortality, was Lovelace also addressing the issues of paternal/maternal identity that not only persisted through her own lifetime, but would colour the way she was viewed after death? Just as the blood of her father, the Romantic, iconoclastic poet, ran through her veins, so too did the mathematical and controlling nature of her mother. Lovelace spent long periods of her life pondering and pursuing the idea of being either a mathematician and a scientist or a poet and a musician, but it was her work with Babbage that allowed her to realise her true potential as a metaphysical analyst. Her mathematical skill enabled her to interpret the underlying mechanics of the Analytical Engine, but it was her poetic insight that enabled her to fully understand that the machine was capable of far more than the calculation of 'simple' equations, thereby paving the way for Alan Turing's work on universal computing more than a hundred years later.

The story of Ada Lovelace is a tragic one; an exceptional talent who died young, and yet who in her short life could see the intersections between poetry and science and use them to create new processes and understanding. Despite her short life and the numerous obstacles that were stacked against her, the poetical science of Lovelace lives on, both in the world of computing and as an advocate for gender equality in STEM subjects. While Byron's poetry remains popular, the ubiquity of the universal computer[48] means that Lovelace's assertion to Babbage that she would be a better analyst than her father had been a poet might now be considered realised. Or, as Lovelace herself put it in another letter to Babbage, dated 5 July 1843: 'That *brain* of mine is something more than merely *mortal*; as time will show.'[49]

3

The lyrical visionary:
James Clerk Maxwell

My life's undivided devotion
  To Science I solemnly vowed,
I'd dredge up the bed of the ocean,
  I'd draw down the spark from the cloud.
To follow my thoughts as they go on,
  Electrodes I'd place in my brain;
Nay, I'd swallow a live entozöon,
  New feelings of life to obtain.

From 'Song of the Cub' by James Clerk Maxwell[1]

Daftie and Richard Goldie

James Clerk Maxwell was born on 13 June 1831 in Edinburgh, Scotland. His father, John Clerk Maxwell, came from a privileged background, having inherited the Middlebie estate near the village of Corsock in the rural region of Dumfries and Galloway (approximately 90 miles south-west of Edinburgh) when his grandmother, Dorothea Clerk Maxwell, died in 1793.[2] Maxwell's mother, Frances, also came from a relatively prestigious background; her father was Robert Hodshon Cay, the Judge Admiral of Scotland, while her mother was the amateur pastel artist Elizabeth Liddell.

Soon after the birth of Maxwell (John and Frances' only son; they had lost an earlier child, Elizabeth, during infancy), the family moved to the Middlebie estate and into the newly built Glenlair House. This is the house that Maxwell grew up in, and to which he would continue to return throughout his adult life. As a young boy, Maxwell spent a lot of time playing with the local children, resulting in a strong Galloway accent that would remain with him for life. He also had a very close relationship with his parents, certainly much closer than would have been the norm for the gentry at the time. These bonds were further nurtured by his accompanying his father on estate business and by the fact that his early education was both organised and delivered entirely by his mother.

As an eight-year-old Maxwell was able to recite the entirety of the 119th Psalm from memory, an impressive feat given that, at 176 verses, it is the longest psalm (and indeed the longest chapter) in the Bible, and that it does not have any rhyming structure or metre that particularly aids recollection.[3] Given what Maxwell would go on to achieve, the second and third couplets of the third stanza of Psalm 119 are particularly pertinent:

> Be good to your servant while I live,
>   that I may obey your word.
> Open my eyes that I may see
>   wonderful things in your law.

The recollection of long passages of work extended beyond scripture and included the works of poets such as Milton and Burns, and it is likely that from an early age Maxwell had a powerful photographic memory. The following statement, taken from an essay written by Maxwell towards the end of his life (in 1878), is indicative of this:

> By means of the instantaneous light of a single electric spark, we may read a whole sentence of print. Here we know that though the illumination lasts for a few millionths of a second, the image on the retina lasts for a time amply sufficient for an expert reader to go over it letter by letter, and even to detect misprints.[4]

The lessons of his mother were complemented by a fascination for the outdoors and a burning desire to determine the way in which all things worked. The young Maxwell could often be heard asking both his parents and the household staff 'What's the go o' that?' followed immediately by 'But what's the particular go of it?' if he was not provided with a suitably in-depth answer. While his photographic memory made it possible for Maxwell to memorise long passages from a variety of texts, it was his tireless quest to work out exactly 'what the go' of everything was that would ultimately prove to be his biggest virtue. According to Maxwell himself, one of his earliest recollections was of simply lying on the grass in front of the family home while looking at the sun and wondering.

Maxwell's mother died of abdominal cancer when he was just eight years old. This event had two immediate impacts: it brought father and son even closer together (despite only being in his fifties when his wife died, Maxwell senior would not marry again), and it necessitated the hiring of a tutor. The latter of these outcomes was, by all accounts, an unequivocal failure.[5] For while under the formal tutelage of his mother, and the informal teachings of his father, young Maxwell had been encouraged to question the world and the way in which it worked, such an approach to education was perhaps too challenging for his tutor, himself a boy of just 16. The new tutor reported his student as being slow at learning, and discouraged Maxwell's more adventurous lines of questioning. Eventually it transpired (via letters between Maxwell and his aunt) that the tutor was beating Maxwell for poor performance, leading to his eventual dismissal. While the tutor had been in post for just over a year, his poor teaching and overly rough treatment were to have a lasting effect on Maxwell, who continued to display a certain hesitation when replying to questions throughout the rest of his life.

The incident with the tutor convinced Maxwell's father that it was time for his son to attend school, and so in November 1841 Maxwell was sent to the celebrated Edinburgh Academy. Unfortunately, things did not begin well here either; his strong Galloway accent was in stark contrast to the Edinburgh brogue of his classmates, while the loose tweed tunic and square-toed shoes (both made by his father and no doubt given to the boy as a parting gift) that he wore on his first day at school clashed heavily with the tight jackets and narrow shoes that were de rigueur among affluent Scottish schoolchildren at the time. His unfortunate sartorial stylings and thick rural accent resulted in Maxwell being given the nickname 'Daftie'; proof if it were needed that young schoolchildren are not always the best judges of character.

Maxwell's life at the Academy improved greatly when he made friends with Lewis Campbell, a boy in the same class who also lived very close to his Aunt Isabella, with whom Maxwell resided during term time (the holidays were spent at Glenlair). Lewis Campbell went on to become a renowned Scottish classical scholar, and the two would remain lifelong friends, with Campbell writing a biography shortly after Maxwell's death, which also contains an almost complete collection of his poetry.[6] Shortly after meeting Campbell, Maxwell made friends with another classmate who would also go on to become a lifelong friend: Peter Guthrie Tait. Alongside Maxwell, Tait would later become one of Scotland's most celebrated physicists, and as can be observed from *The Scientific Letters and Papers of James Clerk Maxwell*, the correspondence between the three demonstrates a deep and sincere friendship that persisted throughout their lives.[7]

At first Maxwell did not particularly excel in his studies, which may in part have been because he had to join the school in a second-year class, the first year being full when his father applied. However, it was not long before Maxwell's aptitude and intellect shone through, and in the second year of his schooling he won the school prize for scripture biography, testament to a knowledge of the Bible that had been nurtured in him since he was a small child. In the third year of his studies his year group were introduced to mathematics, and this is where Maxwell quickly began to outshine all his peers, winning the mathematical medal in the summer of 1845. At the same time, Maxwell also won the English prize for a poem that he wrote commemorating the death of James Douglas, a Scottish knight and one of the chief commanders during the Wars of Scottish Independence. The opening three couplets of the poem, entitled 'The Death of Sir James, Lord of Douglas', read as follows:

Where rich Seville's proud turrets rise
A foreign ship at anchor lies;
The pennons, floating in the air,
Proclaim that one of rank is there —
The Douglas, with a gallant band
Of warriors, seeks the Holy Land.[8]

Maxwell prefaced these lines with the following quote from the Scottish poet John Barbour's long narrative poem 'The Brus':

'Men may weill wyt, thouch nane thaim tell,
How angry for sorow, and how fell,
Is to tyne sic a Lord as he
To thaim that war off hys mengye.'

Barbour's *Bruce*, B. XX. i. 507.

Maxwell was clearly inspired by Barbour, writing to his aunt (Miss Cay) in June 1845 to inform her that he had been studying a great number of books in preparation for his poem about Sir James Douglas, and that in doing so the above quotation had become his motto.

'The Death of Sir James, Lord of Douglas' was, however, not the first poem to be written by Maxwell. The two earliest attempts that we know of are a poem written during the 1843 summer holidays in Glenlair — which is omitted from Campbell's collection of Maxwell's poetry on the grounds that it is unremarkable, other than in having 'characteristic touches of grotesque ingenuity and humorous observation which are very curious in a boy of twelve'[9] — and a short poem written just before 'The Death of Sir James, Lord of Douglas', addressed to his father and concerning the death of a goldfinch. It begins:

Lo! Ossian makes Comala fall and die,
Why should you not for Richard Goldie cry?[10]

Ossian was the narrator of a cycle of epic poems by the Scottish poet James Macpherson, with Comala the eponymous and tragic heroine of one such poem. By asking his father to mourn for a pet goldfinch as if it were the dead heroine of a classic Gaelic tragedy, the author's 'grotesque ingenuity and humorous observation' are again apparent.

In the year after winning the English and mathematics prizes at the Academy, Maxwell's academic abilities developed further, and he successfully published his first academic paper, 'Oval Curves', in the *Proceedings of the Royal Society of Edinburgh* when he was just 14. Given his age, Maxwell was deemed too young to present the paper himself,[11] and so it was instead given by James Forbes, a friend of Maxwell's father and the Professor of Natural Philosophy at the University of Edinburgh. Maxwell senior was himself a member of the Royal Society of Edinburgh and had piqued Forbes's interest by sending him the workings of his son, which Forbes identified as being a simplification and improvement of work done by the French mathematician René Descartes some two hundred years earlier. It was Forbes who would later convince Maxwell's father that his son should study mathematics and not law at university, with Forbes also lecturing Maxwell in natural philosophy when he moved to the University of Edinburgh at the age of 16 in 1847.

Shortly after leaving the Academy, Maxwell wrote two further poems, 'An Onset' and 'The Edinburgh Academician'.[12] The latter of these two is a farewell to his old school, while the former is a call-to-arms which is reminiscent of the heroic odes of Maxwell's poetic heroes of the time, Macpherson and Barbour. The first few lines are as follows:

Hallo ye, my fellows! arise and advance,
See the white-crested waves how they stamp and they dance!
High over the reef there in anger and might,
So wildly we dance to the bloody red fight.
Than gather, now gather, come gather ye all,
Each thing that hath legs and arms, come to our call;
Like reeds on the moor when the whirlwinds vie
Our lances and war-axes darken the sky;[13]

Throughout his life Maxwell was an intensely private person, and it is unlikely that he wrote most of his poetry for the sole purpose of others reading it; as such, the poems that are most auto-biographical in nature provide a possible insight into Maxwell's thoughts and feelings at the time of writing. The youthful exuberance of these poems suggests that Maxwell was in good spirits about his forthcoming studies at the University of Edinburgh. They also bear the frustrations of a student who was aware that he had surpassed the capacities of his former tutors, something that was also observed by Campbell when reminiscing on their time together at the Academy:

> It was thought desirable that we should have lessons in 'Physical Science.' So one of the classical masters gave them out of a textbook. The sixth and seventh[14] classes were taught together; and the only thing I distinctly remember about these hours is that Maxwell and P.G. Tait seemed to know much more about the subject than our teacher did.[15]

Wrangling and dreamland

While Maxwell enjoyed his time at the University of Edinburgh, he did not find any of the classes particularly taxing, and so was able to devote most of his time to developing his own theories and experiments. This period of self-study paid off, as while he was a student at the university, two more of his papers were presented to the Royal Society of Edinburgh: 'On the Theory of Rolling Curves' and 'On the Equilibrium of Elastic Solids'.[16] As with his previous research, Maxwell was still considered too young to present the work himself, and so it was presented on his behalf by of one of his tutors, the English mathematician Philip Kelland.[17]

Maxwell's scientific and poetic outputs while he was studying at the University of Edinburgh are worth contrasting, as while he produced two original and significant contributions to scientific knowledge during this time, he failed to write a single poem.

After 'An Onset' and 'The Edinburgh Academician', Maxwell did not write another poem until 1852. As we shall see later, Maxwell wrote at least one poem a year throughout his entire adolescent and adult life apart from two key periods, one of which was during this time as a student. The original[18] poetry that Maxwell had produced up until this point had been written as an alternative way for him to make sense of the world in which he lived, be it either the death of a goldfinch or his excitement at starting university. I therefore suggest that during these formative years at the University of Edinburgh, Maxwell was so wrapped up in his pursuit of science that he did not have the time or the inclination to write poetry. While his formal studies at the University of Edinburgh appear to have been little more than a distraction, he was surrounded by tutors, particularly Forbes, who had the capacity to challenge and inspire him. Indeed, it was Forbes who gave Maxwell the use of his personal laboratory, encouraging him to develop his experimental acumen in the process. It would appear that while Maxwell was able to pursue scientific interests at his own leisure, he was able to use science to find out what the 'go o' things' were, meaning that he had less need for the alternative explorations that poetry could provide.

In October 1850 Maxwell moved to the University of Cambridge to study for the Mathematical Tripos (the taught mathematics course in the Faculty of Mathematics). Initially he enrolled in Peterhouse College, but after less than a term he transferred to Trinity, as he believed that he had a better chance of obtaining a fellowship there after graduation. While Maxwell continued to excel in his studies at Cambridge, he did not find them as straightforward as those at Edinburgh. At the time the questions in the Tripos rarely bore much resemblance to real-life problems, and in order to answer them all fully, and within the time limit, a variety of mathematical tricks had to be employed, something that contrasted with the methodological approach that Maxwell desired to employ. His frustrations in relation to this period of study are apparent in the opening stanza from 'A Vision of a Wrangler, of a University, of Pedantry, and of Philosophy', dated 10 November 1852:

DEEP St. Mary's bell had sounded,
And the twelve notes gently rounded
Endless chimneys that surrounded
   My abode in Trinity.
(Letter G, Old Court, South Attics),
I shut up my mathematics,
That confounded hydrostatics—
   Sink it in the deepest sea![19]

Wrangler is the unofficial title bestowed upon students who obtain an undergraduate mathematics degree at the University of Cambridge with first-class honours; a Senior Wrangler is the student with the highest graduating mark, the Second Wrangler the student with the second-highest mark and so on. Notable Senior Wranglers from this period included Philip Kelland, Maxwell's tutor at the University of Edinburgh, in 1834, and Peter Guthrie Tait, Maxwell's friend and confidant, in 1852.[20]

A year to the day after his initial protestations (on 10 November 1853), Maxwell wrote another poem about the frustrations of his studies, this one entitled 'Lines written under the conviction that it is not wise to read Mathematics in November after one's fire is out'; the third stanza (of twelve) begins:

Why should wretched Man employ
Years which Nature meant for joy,
Striving vainly to destroy
   Freedom of thought and feeling?[21]

From these lines, it is apparent that Maxwell was frustrated by the Tripos, and that he considered studying for such (at times) esoteric examinations a waste of his time; time that he could otherwise have spent freely investigating scientific pursuits, as he had done in Edinburgh. Yet despite his protestations, Maxwell wanted to perform well in his studies, as evidenced by him joining the tutor group of William Hopkins, who at the time was considered the best 'Wrangler-maker' (i.e. he had coached the highest number of well-ranked Wranglers) in Cambridge. In 1854 Maxwell eventually finished Second Wrangler, runner-up to the English mathematician Edward Routh, who would later go on to be considered one of the most successful Wrangler-makers of all time.[22] The difficulties that Maxwell faced in studying for the Tripos, and the frustrations that can be seen in his poetry, ultimately led to a nervous breakdown after his third-year examinations, which he recovered from thanks to the help of the Revd Charles Benjamin Tayler (the uncle of one of his classmates), with whom he was staying at the time.

Immediately after graduating, Maxwell presented a paper 'On the Transformation of Surfaces by Bending' to the Cambridge Philosophical Society.[23] This was followed a year later by the paper 'Experiments on Colour, as perceived by the Eye, with Remarks on Colour-Blindness', which he presented, this time in person, to the Royal Society of Edinburgh.[24] Already Maxwell was developing a formidable reputation as both a mathematician and a physicist, with expertise across a wide range of topics. This reputation led to him being made a Fellow of Trinity in October 1855, followed swiftly by an application to become the Professor of Natural Philosophy at Marischal College, Aberdeen. Maxwell was helped in the application process by his father, and took up the post in the autumn of 1856 at the age of just 25. Sadly, Maxwell's father did not live to see his son's success, dying on 3 April 1856 before the new appointment was confirmed.

At the time of his father's death, Campbell notes that 'the outward change was not very great' in Maxwell, and this is also apparent from the letters that he sent at the time, most of which were business-like and perfunctory in manner, concerning the upkeep of the Glenlair estate and his father's will.[25] However, given the closeness of the two men, the loss must have affected him deeply, as can be seen in the following lines from 'Recollections of Dreamland', a poem written in April 1856 shortly after his father's death:

Yet my heart is hot within me, for I feel the gentle power
Of the spirits that still love me, waiting for this sacred hour.
Yes, —I know the forms that meet me are but phantoms of the
    brain,

For they walk in mortal bodies, and they have not ceased from
  pain.
Oh! those signs of human weakness, left behind for ever now,
Dearer far to me than glories round a fancied seraph's brow.
Oh! the old familiar voices! Oh! the patient waiting eyes!
Let me live with them in dreamland, while the world in slumber
  lies![26]

His father's death was clearly not something that Maxwell could
resolve through any scientific or theoretical endeavour, and so
instead he turned to poetry, using it as a tool through which to
explore his pain and perhaps better deal with his loss.

Piercing the gloom

Following the death of his father, Maxwell settled into his professorship in Aberdeen. This new role included 15 hours of teaching
per week, which given the additional time needed for preparation
and departmental administration was no doubt a 'considera-
ble burden for a man also intent on doing front-line research'.[27]
However, unlike his studies for the Tripos, Maxwell felt he had
sufficient time for original thought, devoting the first few years
of his time at Aberdeen to investigating the nature of Saturn's
rings. A proof for how they remained in stable orbit around the
planet was the topic of the 1857 Adams Prize (an award from the
University of Cambridge for the most distinguished mathematical
research on a given topic), which Maxwell won in 1859 with his
essay 'On the Stability of the Motion of Saturn's Rings'.[28] In this
essay, Maxwell showed that the rings were in fact made up of many
smaller particles, all of which orbited Saturn in their own individual orbits. Maxwell's theory was later proved to be correct by
spectroscopic observations made by the astronomers James Keeler
and William Campbell in 1895, with further proof coming during
the flybys of the planet by the *Voyager* spacecraft in the early 1980s.
During these flybys, a gap between two of the rings was observed
and christened 'the Maxwell Gap'.

During his work on Saturn's rings, Maxwell married Katherine Mary Dewar, daughter of the then principal of Marischal College, the Revd Daniel Dewar. Little is known about Katherine, other than that she assisted her husband in some of his experiments (particularly the work on colour matching that he undertook while at Aberdeen), and that his marriage was strong but almost invisible to outsiders.[29] The strength of their marriage is evidenced by Campbell,[30] who stated that they lived a 'married life which can only be spoken of as one of unexampled devotion'.[31] This devotion is clear from the final stanza of an untitled poem written in 1858, which was given the title 'To His Wife' by Maxwell's biographers:

All powers of mind, all force of will,
    May lie in dust when we are dead,
But love is ours, and shall be still,
    When earth and seas are fled.[32]

At the time of his work on Saturn's rings and his marriage to Katherine, the only poetry written by Maxwell is what Campbell would class as 'Occasional' verse, most of which was concerned with positive thoughts relating to either nature or his wife. However, while the subject matter of this poetry indicates that Maxwell was content in his personal life (and not facing any obvious barriers in his professional one), that such poetry exists at all is interesting. While it might be expected for a newlywed to write poetry to/about his beloved, the relative abundance of poetry that was written in this period suggests that Maxwell was perhaps not as fully scientifically occupied as he had been at Edinburgh, and as he would go on to become in the next period of his life.

In 1860 Marischal College merged with nearby King's College to form the University of Aberdeen. As there was only one Chair in Natural Philosophy available, Maxwell found himself in competition with David Thomson, who held the position at King's. Despite his burgeoning reputation, and the position of his father-in-law, Maxwell lost out to Thompson, a turn of events that is now attributed to the 'crafty' nature of Maxwell's competitor and his great influence within the commission that was tasked with the unification process.[33] At the same time, Maxwell was also unsuccessful in his application to take up the Chair of Natural Philosophy at the University of Edinburgh (vacated by his old tutor James Forbes, who had left to become principal of the United College of St Andrews in 1859), losing out to Tait, his close friend and compatriot. Maxwell was eventually successful in an application for a new position, taking up the Chair of Natural Philosophy at King's College, London and moving there in 1860.

Upon arrival at King's College, Maxwell continued with the work on colour that he had begun in Edinburgh and Cambridge, and was awarded the Rumford Medal of the Royal Society for his work.[34] Shortly after this award, in 1861, Maxwell was elected a Fellow of the Royal Society, and later in that same year he collaborated with the English inventor Thomas Sutton to produce the world's first colour photograph, demonstrating the process in a lecture that he gave for the Royal Institution. During his time at King's College, Maxwell also began the research into electromagnetism for which he is best known, particularly his work in determining the underlying nature of both electric and magnetic fields and demonstrating how light propagates as electromagnetic waves.

At the beginning of the 1800s several experimental observations and numerical formulations by the likes of the French engineer and physicist Charles-Augustin de Coulomb and the German mathematician and physicist Carl Friedrich Gauss had resulted in a series of equations that represented different aspects of electricity and magnetism, albeit with each of the forces being represented independently of one another. This changed in 1831 when Michael Faraday discovered electromagnetic induction, a process by which an electromotive force is produced across an electrical conductor because of the way that the conductor interacts with the surrounding magnetic field. Faraday explained his discovery via lines of force, although as he was unable to mathematically prove their existence they were not readily accepted by other scientists. Through a series of papers published in the early 1860s Maxwell *was* able to mathematically describe this relationship, producing a set of equations that have come to be known as 'Maxwell's equations'. The history of Maxwell's equations is a somewhat nuanced affair, as while the four modern equations that are today taught to undergraduate scientists the world over do appear individually throughout Maxwell's 1861 paper 'On Physical Lines of Force', it was actually the English mathematician and physicist Oliver Heaviside who grouped together all of the original twenty equations that Maxwell suggested, presenting them as the four modern equations via the use of vector notation.[35]

In addition to his mathematical formalisation of the relationships between electricity and magnetism, Maxwell also demonstrated that light is in fact an electromagnetic wave, stating in 'A Dynamical Theory of the Electromagnetic Field' that 'The agreement of the results seems to show that light and magnetism are affections of the same substance, and that light is an electromagnetic disturbance propagated through the field according to electromagnetic laws.'[36] The profundity of this work should not be understated, with Richard Feynman later observing that 'From a long view of the history of mankind – seen from, say, ten thousand years from now – there can be little doubt that the most significant event of the 19th century will be judged as Maxwell's discovery of the laws of electrodynamics.'[37]

In 1865 Maxwell left his post at King's College and returned to Glenlair to continue his work on electromagnetism, thermodynamics, and the perception of colour. There is some debate as to whether Maxwell left his professorship voluntarily or whether he was asked to leave because of his shortcomings as a teacher; an additional lecturer was certainly appointed to assist Maxwell in his teaching duties, an act that might ultimately have brought about his departure.[38]

Throughout the construction of Maxwell's equations there are no recorded instances of any poetry, with the most recent poem relative to this period being 'To His Wife'. In a similar vein to his time at Edinburgh, the 1860s seemed to have presented Maxwell with sufficient opportunities to pursue his scientific interests relatively freely, surrounded by other scientific luminaries who inspired him (particularly Faraday). If indeed he was pushed from his position at King's, Maxwell did not appear to bear any grudges, perhaps because, free of any lecturing commitments (and not reliant upon a teaching salary given his large estate), he was permitted even greater freedom to continue to make sense of the world in a manner that suited him best. The lack of poetry during this period, and the subject matter of his previous efforts, lends further credence to the notion that Maxwell wrote poetry to make sense of the world around him, but only when his scientific endeavours did not provide the answers, or else when there was something/someone obstructing him in such pursuits. Maxwell's poetic silence was to be broken in the succeeding decade, when in 1871 he took up an offer from the University of Cambridge to become the first Cavendish Professor of Experimental Physics.

Reflections and materialism

The funds for the new Cavendish Laboratory were met by the then chancellor of the university, William Cavendish, the Seventh Duke of Devonshire and relative of Henry Cavendish, who agreed to pay for the laboratory on the proviso that the university colleges funded a new Chair in Experimental Physics, the position that was given to Maxwell. Henry Cavendish was an English scientist who is perhaps best known for his discovery of hydrogen, and his famous experiment in which he measured the mass of the Earth (and with it the gravitational constant) by observing the faint gravitational attraction between two massive lead balls and two smaller ones. In becoming the first Cavendish Professor of Physics, Maxwell performed two notable tasks: he edited the original notes and research of Cavendish, and he established the Cavendish Laboratory, of which he was the founding director. By thoroughly examining Cavendish's notes and lab books, Maxwell revealed that Cavendish had made several scientific discoveries that were later deduced by other scientists. Such discoveries included Ohm's and Coulomb's laws, but until Maxwell came along they remained hidden, mainly because of Cavendish's reluctance to publish, behaviour that ret-rospective research has in part attributed to a suspected form of Asperger's syndrome.[39]

When Maxwell began setting up the Cavendish Laboratory, many thought that the conservative approach of the university (which Maxwell had come up against when studying for the Tripos) precluded the conditions necessary for meaningful physics (and science in general) to be conducted there, with physics still considered by many at the university to be a purely theoretical pursuit. Shortly after Maxwell's appointment, Joseph Norman Lockyer wrote in his capacity as editor of the influential science journal *Nature* that 'Despite Maxwell it may take Cambridge thirty to forty years to reach the level of a second-rate German university in physical research.'[40] During the five years of Maxwell's direction Lockyer's words proved to be unfounded, as the laboratory and the scientists who worked there flourished. This was largely due to Maxwell's attitude towards conducting scientific research, summed up in the following statement that he made to a colleague during that period: 'I never try to dissuade a man from trying an experi-ment. If he does not find what he wants, he may find out something else.'[41]

Maxwell's own experimental skills had been developed under Forbes, and further honed in the personal laboratory that he had set up at Glenlair. He brought this systematic approach to Cambridge, and in doing so also had the good sense and authority to simply let scientists get on with their work, providing them with the equip-ment that they needed, as well as assistance and feedback when required. The Cavendish Laboratory has since gone on to produce twenty-nine Nobel Laureates, including J. J. Thompson in 1906 for his work on the conduction of electricity in gases, and Francis Crick and James Watson in 1962 for developing a structural model of DNA, none of which would have been possible without the legacy laid down by Maxwell.[42] During his time as director of the Cavendish Laboratory, Maxwell also began writing poetry again. As we shall see, this poetry was written as a defence of the way in which he believed science should be conducted, a process that was intrinsically linked to his belief in the relationship between science and religion.

While Maxwell had been raised as both a Presbyterian (by his father) and an Anglican (by his aunt), it wasn't until he was a student at Cambridge that he began to develop the evangelicalism that would remain with him throughout the rest of his life. It was during this time that he began to give serious consideration to a number of religious questions, but found that he was unable to answer them using the empirical approach with which he conducted his scientific investigations. Instead, poetry provided him with the space in which to conduct these spiritual explorations. This can be seen most clearly in 'Reflex Musings: Reflections from various Surfaces', written in April 1853 following Maxwell's inability to fully understand the origin of evil;[43] the third stanza of this poem reads as follows:

> By the hollow mountain-side
>> Questions strange I shout for ever,
> While the echoes far and wide
>> Seem to mock my vain endeavour;
>> Still I shout, for though they never
> Cast my borrowed voice aside,
>> Words from empty words they sever –
> Words of Truth from words of Pride.[44]

By the summer of 1853 Maxwell appears to have resolved these questions and converted fully to evangelicalism. This conversion is largely thought to have been influenced by the breakdown that occurred following his third-year examinations, and the subsequent care and affection that Maxwell received from the Taylers as they nursed him back to full health.

During the Victorian era of science that Maxwell inhabited there was an ongoing and heated debate about the role of science and religion. Following on from Charles Darwin's revolutionary research into evolution, scientists such as the English biologist Thomas Henry Huxley and the Irish physicist John Tyndall worked hard to popularise scientific materialism. This is the belief that physical reality, as proven by empirical (i.e. measurable) tests, is all that truly exists; a position that has little room for religion, which involves faith in the unseen and does not require empirical tests in order to be believed. The views of the materialists were in stark contrast to those of other scientists of the period (such as Tait and Faraday), who believed that religion and science were agreeable and congruous with one another.

In contrast, Maxwell declared several times that science had nothing to say on matters of religion; it wasn't so much that they were compatible or incompatible as that they were utterly incomparable, and he once wrote to the bishop of Gloucester and Bristol, advising against the 'misuse of partial scientific knowledge to interpret scripture, let alone to shore up faith by supposed harmonization with the latest science'.[45] However, on occasion he did make public reference to religious matters, such as his opinions with regard to 'molecular perfection'. Maxwell believed that the perfectly identical nature of molecules implied that they must have been created according to some 'intelligent design', pointing to the ordered uniformity of nature as signs of a creator and not as something that could be entirely explained by the evolutionary adaptation outlined by Darwin and championed by Huxley and Tyndall. This belief is neatly summarised in the conclusion to Maxwell's 'Molecules' lecture, which was delivered to the British Association for the Advancement of Science (BAAS) at its 1873 meeting in Bradford and then published in *Nature*:

> They continue this day as they were created – perfect in number and measure and weight, and from the ineffaceable characters impressed on them we may learn that those aspirations after accuracy in measurement, truth in statement, and justice in action, which we reckon among our noblest attributes as men, are ours because they are, essential constituents of the image of Him who in the beginning created, not only the heaven and the earth, but the materials of which heaven and earth consist.[46]

This molecular perfection argument was critiqued by Tyndall in his presidential address to the 1874 annual meeting of the BAAS,

held in Belfast at the Ulster Hall.[47] In this address Tyndall also outlined his intention for the future of science, that is, that materialism
should form its true philosophy from that point forward. While this
might be viewed as an attack on religion (which Tyndall and the
other BAAS materialists argued could not be empirically 'proved'
and was therefore not 'real'), it potentially had other, far-reaching
consequences. For if such materialism were taken to the extremes
that were being proposed, then there would be no need for any of
the humanities, or indeed any philosophical ideology that could
not be empirically tested. Maxwell saw the inherent danger in this,
responding brilliantly with a poem entitled 'British Association,
Notes of the President's Address', which was subsequently published in *Blackwood's Edinburgh Magazine*. The following section
of the poem clearly highlights Maxwell's derision at the notion
that all feelings and emotions can be boiled down to nothing more
than the behaviour of atoms, while poking fun at Tyndall's Belfast
address:

> Thus in atoms a simple collision excites a sensational thrill,
> Evolved through all sorts of emotion, as sense, understanding,
>   and will;
> (For by laying their heads all together, the atoms, as councillors
>   do,
> May combine to express an opinion to every one of them new).
> There is nobody here, I should say, has felt true indignation at all,
> Till an indignation meeting is held in the Ulster Hall.[48]

The arguments that Maxwell makes in this poem reinforce the
point of view that he had already underlined in his 'Molecules'
lecture, in which he clearly states that statistics is not the only way
to make sense of the world in which we live:

> This [statistics], of course, is not the only method of studying
> human nature. We may observe the conduct of individual men and
> compare it with that conduct which their previous character and
> their present circumstances, according to the best existing theory,
> would lead us to expect.[49]

Following Tyndall's presidential address, Tait co-wrote a book, *The Unseen Universe*, with another Scottish physicist, Balfour Stewart.[50] *The Unseen Universe* was written as a direct response to Tyndall's materialism, and argued that religious concepts such as immortality and miracles were in fact compatible with modern science. However, just as he had done with the bishop of Gloucester and Bristol, Maxwell was quick to point out that such an approach was unnecessary; religion did not need to be empirically tested, nor did it have to answer to science. Maxwell's responses to his religious and academic colleagues offer further proof that his issues with Tyndall were not so much with his views on religion, as his materialist assertions that everything must be 'empirically testable', and that such an approach was the blueprint that should be adopted for the future philosophy of science.

Song of the cub

By the time of Tyndall's Belfast speech it was typical for the presidential address of the BAAS to be satirised by a group of BAAS members who referred to themselves as the Red Lion Club. This club was formed at the 1839 BAAS annual meeting, when a breakaway group of young scientists, tired of the pomp and pageantry of the meeting's formal dinner, decamped to a local pub (The Red Lion) to dine on beef and beer instead. These meetings continued annually, with a bohemian attitude that served to lampoon the formalities of the BAAS, while also making a more serious statement that associations for scientists should primarily be composed of scientists; as discussed in Chapter 1, several of the supposedly scientific societies had large proportions of non-scientific members from high society, something that did not sit well with scientists such as Huxley and Tyndall, who were seeking to professionalise science. In a manner that was analogous to the hierarchies of the BAAS, the Red Lion Club was presided over by a 'Lion King', ordinary members called 'cubs', and the 'jackals' who organised the meetings, including the catering.[51] At these dinners various parodies of BAAS lectures would take place, either via the comic recreation of experiments or the satirising of the speeches. Maxwell contributed several such parodies, including the 'British Association, Notes of the President's Address', and also the following poem, 'Song of the Cub', which was written at the same time:

I know not what this may betoken,
    That I feel so wondrous wise;
My dream of existence is broken
    Since science has opened my eyes.
At the British Association
    I heard the President's speech,
And the methods and facts of creation
    Seemed suddenly placed in my reach.

My life's undivided devotion
    To Science I solemnly vowed,
I'd dredge up the bed of the ocean,
    I'd draw down the spark from the cloud.
To follow my thoughts as they go on,
    Electrodes I'd place in my brain;
Nay, I'd swallow a live entozöon,
    New feelings of life to obtain.

O where are those high feasts of Science?
    O where are those words of the wise?
I hear but the roar of Red Lions,
    I eat what their Jackal supplies.
I meant to be so scientific,
    But science seems turned into fun;
And this, with his roaring terrific.
    That old red lion hath done.[52]

Maxwell was concerned with the over-burdening and dogmatic approach that was being proposed by Tyndall in his address; a stark contrast to the relatively laissez-faire attitude with which he encouraged scientists to work in the Cavendish Laboratory. And Maxwell satirises and pokes fun at this idea in 'Song of the Cub'; in the first stanza, he mocks that there is no need for religion now that

Tyndall has explained every facet of nature (or at least intended to do so) with scientific materialism. The absurdity of the second verse also highlights the seriousness of Maxwell's assertions. Like the fictional and fanatical evangelist that he presents, Maxwell has also devoted his entire life to science, and while he hasn't done anything as physically dramatic as swallowing a 'live entozöon',[53] he *has* dedicated his life to the advancement of scientific knowledge. Given the privilege of his upbringing and the comfort that his estate afforded him, Maxwell had absolutely no need to practise science, other than his desire to make sense of the world in which he lived. Given that Maxwell believed in an incomparable relationship between science and religion (something that the poetry that he wrote while a student at Cambridge had helped him come to terms with), he could dedicate his life to finding the 'Words of Truth' without calling into question the devotion to his evangelical beliefs. Thus, in writing this second stanza, Maxwell is exposing the fallacy of Tyndall's extreme materialism, but he is also laying bare the degree to which he has already worshipped at the altar of science, thereby highlighting the extent to which he is aggrieved by Tyndall's comments.

In the third stanza, Maxwell continues with the religious allegories; if science is to be the new religion, then does Tyndall really expect people to sit down and celebrate key scientific achievements in a manner akin to religious high feasts? Rather than wise words that will help to positively shape their followers, Maxwell remarks that the only sound that can be heard is the roar of the Red Lions. The emphasis of these meetings was on 'horse-play and jocular fraternity';[54] so, when Maxwell declares in his poem that 'science seems turned into fun', he is perhaps pointing out the ridiculousness of Tyndall in claiming that he and other members of this fraternity should be solely tasked with attempting to understand the mysteries of the universe; that anyone who had taken part in such activities as roaring whenever they addressed their compatriots should be held in the same light as credible religious figures he finds to be preposterous. While potentially implicating his own culpability in not challenging such behaviour sooner ('I eat what their Jackal supplies'), Maxwell is clear that it is the old Red Lion, that is, Tyndall, who should be held accountable for the inherent dangers of this new vision of science that he is suggesting.

That Maxwell gave an alternative title for 'Song of the Cub' as 'Molecular Evolution'[55] serves as further evidence that the poem was written as a response to Tyndall's doctrine of scientific materialism. Furthermore, 'Song of the Cub' is written with alternating feminine and masculine endings, that is, the lines alternate between finishing with an unstressed syllable and a stressed syllable. This type of rhyming structure is very popular in traditional hymns. Maxwell was a regular churchgoer for almost all of his life, and was also a great appreciator of poetry;[56] it therefore seems reasonable to assume that he was familiar with hymnal rhyming schemes and that he chose this particular scheme to further ridicule the notion that science and religion had anything to do with one another, thereby satirising Tyndall's own stance in the process.

Poetry and synthesis

Maxwell died in Cambridge of abdominal cancer on 5 November 1879 at the age of 48.[57] He was buried in Parton Kirk, Galloway, close to Glenlair, and was laid to rest alongside his parents; a simple gravestone marks their graves, with Katherine also buried there following her death in 1886.

As has previously been discussed, Maxwell's contribution to science should not be understated; he was arguably the most important scientist of the nineteenth century, and his research into thermodynamics, electromagnetism, and kinetic motion have underpinned the development of these disciplines right through to modern times. With regard to the legacy of his scientific achievements there is almost universal agreement as to its importance and brilliance. However, the appreciation of Maxwell's poetry is not quite so universal. In the preface to *The Scientific Papers of Maxwell Clerk Maxwell*, the editor, W. D. Niven, notes:

> Maxwell was not only a lover of poetry but himself a poet, as the fine pieces gathered together by Mr Campbell abundantly tes-

tify. He saw however that his true calling was Science and never regarded these poetical efforts as other than mere pastime.[58]

However, while it is true that Maxwell did not consider himself to be a poet in the same manner as his literary idols, the process of writing seems to have been far more than an idle pastime. As has been evidenced in this chapter, Maxwell wrote poetry as an alternative to the empiricism of his science. The intervals during which he did not write poetry corresponded to his most prolific periods of scientific achievement, periods in which he was given sufficient freedom to continue his quest to find out 'the go o' things'. It was when such explorations met with frustrations and hurdles (whether the esotericism of the Tripos or the dogmatism of the materialists) that Maxwell turned to poetry, as he also did to explore the more personal aspects of his life (whether the death of his father or his marriage to Katherine). This explanation for why Maxwell wrote poetry is supported by his great friend Lewis Campbell, who in the preface to his biography of Maxwell states that:

> Whatever may be the judgment of critics as to the literary merits of Maxwell's occasional writings in verse, there can be no doubt of their value for the purpose of the present work. Like everything which he did, they are characteristic of him, and some of them have a curious biographical interest. Maxwell was singularly reserved in common life, but would sometimes in solitude express his deepest feelings in a copy of verses which he would afterwards silently communicate to a friend.[59]

The quality of Maxwell's poetry is somewhat undersold by this comment. However, his other lifelong friend Tait (with whom he also shared many of his poems) had a different observation to make in writing Maxwell's obituary: 'No living man has shown a greater power of condensing the whole substance of a question into a few clear and compact sentences than Maxwell exhibits in his verses.'[60] While this statement contains a degree of hyperbole that should perhaps be expected from an obituary written about a close companion, it raises interesting parallels between Maxwell's poetry and his most famous scientific legacy – the Maxwell equations. In deriving these equations, Maxwell summarised all previous understanding of electricity and magnetism, and in so doing revealed new insights that would greatly benefit scientific research and discovery. In reading Maxwell's poetry, we observe an intellect and wit that expressed complex and multifaceted thoughts in clear and succinct verse. In particular, the 'serio-comic' verses, as Campbell would have them, argue succinctly and eloquently against some of the barriers that prevented (or tried to prevent) Maxwell and other scientists from freely practising science. By reading these poems today we can gain an insight into the mind of this great scientist and observe first-hand some of the trials and tribulations that he underwent in assuring his scientific legacy.

4

The medical metrist: Ronald Ross

With tears and toiling breath,
I find thy cunning seeds,
O million-murdering Death.
I know this little thing
A myriad men will save.

From 'Reply' by Ronald Ross[1]

With book on knee and wandering eyes

Ronald Ross was born on 13 May 1857 in Almora, India, in the
Kumaon Hills at the foothills of the Himalayas, to Campbell Claye
Grant Ross and Matilda Charlotte Elderton. Ross's father (like his
father before him) was a soldier, having joined the British Indian
Army as a standard bearer in the 66th Bengal Native Infantry.
His parents married in London during the summer of 1856 while
Campbell was on furlough,[2] returning to India at the beginning of
January 1857. By the time of Ross's birth later that year his father
had risen to the rank of captain and was second-in-command of the
66th Regiment, and he would go on to have a distinguished military career, eventually rising to the rank of general and being made
Knight Commander of the Order of the Bath. Very little is written
of Ross's mother, other than that she was well loved by her family
and that she had ten children, nine of whom survived to adulthood.

Ross was born in tumultuous times; three days before his birth
an uprising against the British East India Company[3] and the
British Crown, which would later become known as the 'Indian

Rebellion', broke out. His father first heard of the mutiny on 15 May 1857, when a couple of Gurkhas warned him that some of the Indian artillerymen stationed with their regiment planned to mutiny and massacre the British officers and their families. Ross's father passed this information on to his commanding officer, Colonel McCausland, who initially waved away the concerns. However, when other Gurkhas made similar reports a few days later McCausland was persuaded to act, and the traitorous artillerymen were surrounded and ejected from the regiment, thereby saving the lives of Ross and his family.

Aside from warfare there were many other dangers for a young baby in India at that time. At the age of three Ross almost died of dysentery, and most of his early childhood was spent following his father's regiment around India, including a trek at the age of seven from Peshawar (which was then in northern India and is now in north-west Pakistan) to Benares (also known as Varanasi), a journey of around 1,600 km that took several weeks. In April 1865 Ross was sent to school in England, making the journey on board a sailing ship called the *Lady Melville*. During this voyage he was put under the care of his uncle, Captain Barwell, a strict disciplinarian who is fondly recalled by Ross in his autobiography: '[Captain Barwell] cured me of several nervous tricks, due to illness, by punishing me when I gave way to them – the proper way to treat such and other weaknesses.'[4]

After arriving in England, Ross was put into the custody of another uncle, William Wilmot, who having practised medicine in Australia for several years, now lived in Ryde on the Isle of Wight. Wilmot, whose first wife had died, was now retired and was remarried to Ross's great aunt, Harriet Elderton. The two of them lived with Harriet's sister, Emma, who was unmarried, and Ross enjoyed their company, especially that of Harriet, whom he viewed as a second mother, later noting in his autobiography that: 'For my aunt Harriet I had the greatest affection, for she became my second mother. Calm, wise, and even learned, she set me the best example of life; and she painted beautiful delicate landscapes, chiefly of Westmorland.'[5] He continued this praise in several of his later poetic verses, some of which appear in the first edition of *In Exile*, including:

Yet, aged? She who loved
And taught – the maxim sage
Lived while her finger moved
About the painted page –[6]

The first school that Ross attended was a small dame school[7] in Ryde, where he was taught in a class of two. By this time he was able to read well, with the first books that he read (after the Bible) being the Elizabethan dramatists – Shakespeare, Chapman, Massinger, and Marlowe. Ross became so imbued with their worlds that he adopted their vernacular and metre, causing his uncle to one day remark: 'Why, the boy talks Elizabethan English!'[8] While the books in his uncle's library were instilling in Ross a literary fascination, the sea anemones, crabs, and minnows of his aunt Emma's aquarium were awakening in him an interest in zoology. This interest continued when he was sent to boarding school at Springhill, near Southampton; here the young Ross was given a small plot of land in the school's gardens, where he kept lizards, frogs, and the occasional snake in a small greenhouse. During this time he also collected butterflies, learning to raise them from caterpillars that he had collected from the surrounding gardens; he even nurtured a chameleon, a gift from his Uncle Charles, which sadly died when one of the gardeners watered it as a cruel joke. In his autobiography (first published in 1923), Ross declared that he was still angry at the gardener for this prank, despite it having occurred almost half a century before; this is emblematic of the defensive nature and feelings of indignation that would later arise in the adult Ross.

During his time at Springhill, Ross quickly took to all of his studies, including grammar, maths, and painting. However, he later considered that his schooling had been uninspiring and formulaic, that the money his parents spent on his education was largely wasted, and that his time would have been better spent teaching himself from the books in his Uncle Wilmot's library. Later in his life, Ross would write two bitingly satirical poems, 'English Spelling' and 'Our Pronunciation of Greek and Latin', both of which were published in the first edition of *Fables and Satires*.[9] 'English Spelling' contains the following verses:

> For speech and spelling don't agree –
> Like country curates after tea.

> We mispronounce our "e's" and "a's"
> And write each sound a dozen ways;
> Altho' we speak a living tongue
> Must learn to write a fancied one.[10]

Similarly, 'Our Pronunciation of Greek and Latin' begins as follows:

> See now the little wretched scholars
> With inky thumbs and irksome collars.

> With fidgeting feet and wand'ring looks
> Attempt to con their blotted books,

> To learn to read and write a tongue
> Which no one ever said or sung;[11]

While Ross's biographer, the English writer and critic Rodolphe Louis Mégroz, later described these poems as reflecting Ross's search for the 'abstract truth more than with a personal resentment', it is difficult to read them without also observing an obvious contempt for his schooling.[12] In particular it is clear that Ross was disdainful about having to learn languages by rote, and was clearly unhappy about being taught a 'living' language (English) in a similar manner to those which he considered 'dead' (i.e. Ancient Greek and Latin).

His parents returned home to England for two extended periods of furlough (also taking up residence in Ryde) while Ross was at Springhill, and he enjoyed spending the holidays with them. In

particular Ross enjoyed watching his father paint the landscapes of
the Indian plains, Kashmir, and the Himalayas, spending hours at a
time observing as he meticulously honed his technique. Ross considered his father to be a 'poet in water-colour', and would often
sit and watch him as he painted in the morning after breakfast.[13]
While his father mainly painted for personal pleasure and rarely
showed his work, Ross learned a great deal from watching him,
and attributed to this his first place, at the age of 16, in the Oxford
and Cambridge local examinations for drawing, with his pencil
copy of what he describes as 'Raphael's Torchbearer'.[14]

As well as being a talented painter, Ross's father was also a fan of
Shakespeare and Byron, the latter of whom he considered a genius,
often quoting him during mealtimes; a favourite passage (perhaps
because its sentiment reminded him of the Indian landscapes of
which he was so fond) was from Act 1, Scene 1 of Byron's dramatic
poem *Manfred*:

Mont Blanc is the monarch of mountains:
They crowned him long ago
On a throne of rocks, in a robe of clouds,
With a diadem of snow.[15]

Despite his love of Byron, and his admiration for other poets
such as Scott and Coleridge, Ross's father was oddly disdainful of
Wordsworth, disparaging his tendency for bathos. It was around
the end of his schooling at Springhill that Ross himself first started
to experiment with poetry, although he describes these earliest
efforts (written on the backs of old letters and in decayed notebooks) as 'painful' and 'full of despair'.[16]

At the end of 1874 Ross left school and decided to become an
artist; however, his father did not approve. Given his passion for
the subject, this disapproval might appear odd, although while his
father was a talented painter, especially with watercolour landscapes, he never considered it as anything other than a hobby.
What is perhaps more unexpected is that Ross's father was also
unimpressed with his son's second choice of career: enlistment in

either the army or the navy. Perhaps he simply didn't want his son to suffer the same alienation from family life that he had no doubt experienced during his time in the army.[17] Ross's father thought that a more admirable and suitable profession for his son would be to become a doctor in the Indian Medical Service. And so, in 1874, shortly before returning to India, he contacted Norman Moore, the Warden of the College at St Bartholomew's Hospital in London, and arranged for his son's apprenticeship into the medical profession. For the 17-year-old Ross this was a career that he had no great interest in, and a route that he found objectionable; ideally, he would have preferred to go either to the University of Oxford or the University of Cambridge before entering the hospital, but Moore had advised Ross's father that this would not be necessary.

While at St Bartholomew's, Ross continued to abhor the rote learning that subjects such as anatomy required, and instead dedicated his time to mastering the piano, composing music, and learning the flute. His interest in poetry was rekindled one summer holiday when he was introduced to the English poet Martin Farquhar Tupper, who was an old friend of the family. After speaking with Tupper, Ross returned to the poetry of his schooldays, and while he kept his poetry hidden from his medical associates (for fear of ridicule), he wrote several poems during this period, some of which would appear many years later in various collections. Of the poems written in this period, perhaps 'The Night Ride' is the most interesting, the opening stanza of which is as follows:

> Who rides by the night in the starlight frore[18]?
> A father, with his child before.
> Oh father, dear father why ride you at night?
> And why do you ride with your sword by your side,
> When all the small stars give no light?[19]

This ballad tells the story of a young boy and his father riding at night along the coast, fleeing an unseen assailant. The identity and nature of the pursuer are left to the reader's imagination, but the poem ends with the horse running off the cliff before they can be caught. Several of the themes that arise in this poem are further developed in Ross's later novel, *The Spirit of Storm*.[20]

When 'The Night Ride' was first published in the 1928 collection *Poems*, Ross found himself accused of plagiarising Goethe's 1782 poem 'Erlkönig'. In this poem, a father is riding on a horse in the dead of night, trying to get his young son to a place of safety where he can receive medical assistance. During the ride, the son keeps seeing the titular Erlkönig, or king of the fairies, while his father reassures him that it is just the darkness playing tricks on him; when they arrive at their destination the son has died. The poem, which is arguably Goethe's most famous ballad, was made more famous by Franz Schubert's 1815 setting. Its opening stanza, translated into English, is as follows:

> Who rides there so late through the night dark and drear?
> The father it is, with his infant so dear;
> He holdeth the boy tightly clasp'd in his arm,
> He holdeth him safely, he keepeth him warm.[21]

While the ballads differ in both their pursuers and denouement, there are distinct similarities between the two pieces, not least in the opening couplet. At the very least, it would be fair to surmise that 'The Night Ride' was inspired by 'Erlkönig'. However, Ross dismissed this, stating in his autobiography that some time in 1878–79:

> I also polished previous verses, and wrote a ballad 'The Night-ride'. The latter is accused of being like the 'Erl-King', but is really quite different – written some years before I read Goethe (though I may have known Schubert's song).[22]

At first this explanation appears reasonable. Unaware of 'Erlkönig', yet a lover of music, it might well have been that the words, themes, and rhythms of Schubert's version bored their way into Ross's subconscious, to be retrieved at a later date. However, we know from Ross's sketchbooks that even from an early age he had been drawn to Goethe, depicting his *Faust* in a sketchbook from 1872 when he was still a teenager.[23] It therefore seems more reasonable to assume that Ross's protestations on the subject are unreliable. While this may seem a trivial point (after all, 'The Night Ride' and 'Erlkönig' are still markedly different poems), Ross's unwillingness to at least acknowledge Goethe as a source of inspiration has direct parallels with the controversies that would later undermine his greatest scientific achievements.

During this period of artistic procrastination at St Bartholomew's, Ross was also fond of visiting one of his uncles, William Alexander Ross. This uncle was a major in the British Indian Army who had taken early retirement, using the small pension that his military career had provided to set up a laboratory in the basement of his house in Shepherd's Bush, London. Here he investigated blowpipe chemistry, a branch of chemistry that involves blowing into a tube to direct a jet of air into a fire or flame. Used since ancient times for soldering metals and working glass, the blowpipe was developed into an analytical tool by the eighteenth-century Swedish chemist Axel Fredrik Cronstedt, who used it to help distinguish between different minerals. Up until the twentieth century, blowpipe chemistry was regularly used by analytical chemists, including Humphry Davy, and Ross's uncle spent a great deal of time (and most of his wealth) in trying to improve the process. He went on to publish several books, which were largely unread, but whose concepts were greatly appreciated by Ross: 'As a matter of fact, nearly all the ideas in science are provided by amateurs, such as my uncle Ross; the other gentlemen write the textbooks and obtain the professorships.'[24]

By 1879 Ross's father was anxious that his son should pass his medical examinations and enter the Indian Medical Service before he himself retired. Despite his lack of enthusiasm for the subject, and with only three days' preparation, Ross somehow managed to pass his first set of examinations, the MRCS (Membership of the Royal Colleges of Surgeons of Great Britain). However, he found himself a victim of hubris, as shortly afterwards (and with even less preparation time) he failed the LSA (Licentiate of the Society of Apothecaries) exams, meaning that he was unable to obtain the full medical qualification required by the Indian Medical Service.[25]

His father was outraged at this failure and threatened to cut off his allowance. Ross therefore took matters into his own hands, stopping his allowance and taking a job as a surgeon on board the steamship *Alsatia*, a moderate-sized vessel running between New York and London; at that time a single qualification, such as the MRCS, was sufficient for this type of employment. As it was no longer possible for Ross to pass his exams and reach India before his father's retirement, he spent his time on board (he made four or five voyages) studying for the LSA and working on his writing. As well as poetry he commenced work on two novels: *The Emigrants*, which described the emigrants to New York that accompanied the *Alsatia*'s cargo, and *The Major*, which described Ross's time as a medical student. Neither of these novels were ever finished.

In 1880 Ross's father (who was by now both a Knight Commander and a general) retired to England, settling with his family in Southampton, where Ross joined them to continue with his revisions. In early 1881 Ross finally passed the LSA, and while he also passed the entrance exams for the Indian Medical Service, he performed poorly (placing seventeenth), meaning that he would have less choice of location for his initial assignment. He had originally wanted to be based in Bengal, and so worked hard in the four months of prescribed military medicine and surgery training that took place at the Royal Victoria Hospital prior to his departure. However, his hard work was unrewarded, and he dropped a further two places in the rankings, giving him even fewer options. At the end of this training he was given six weeks of leave, which he spent reworking his earlier poetry and joining his father for morning painting sessions, watching as he painted the distant Indian landscapes to which his son was about to return. Finally, on 22 September 1881, Ross said goodbye to his family, setting sail for India onboard the troopship *Jumna*.

From a lonely watchtower of the East

After arriving in Bombay,[26] Ross made his way to Madras[27] – the location assigned to him following his poor performance in the

Indian Medical Service examinations. Upon arrival, Ross found that there were no positions in the Indian regiments for new recruits, and so he was instead posted to the Station Hospital, which was exclusively reserved for the treatment of British troops. Given spacious rooms at the nearby Dent's Gardens Hotel, Ross quickly realised that his duties at the Station Hospital were extremely straightforward, requiring only a couple of hours in the morning and the occasional evening visit, thereby giving him both the independence and leisure time to dedicate to his other pursuits. One of his first acts was to visit Higginbotham's bookshop in order to stock up on literature, whereupon, with the help of several dictionaries and grammatical textbooks, he set about working his way through the poets of the world, from Alfieri and Aeschylus to Racine and Virgil.

As well as consuming a huge amount of poetry, Ross also studied for the lower standard examination in Hindustani, the *lingua franca* of northern India. He passed the exams in January 1882; yet despite learning the language, Ross believed himself and other Europeans to be superior to the Indians, considering them to be:

As hard-working as any, faithful, docile, and intelligent, yet in many parts they possess amazingly delicate physique combined with great timidity and a habit of unquestioning obedience; and the swarming millions of them are generally very poor and ill-fed.[28]

The following poem, written by Ross shortly after arriving in Madras and simply entitled 'India', continues this theme:

Here from my lonely watch-tower of the East
    An ancient race outworn I see --
With dread, my own dear distant Country, lest
    The same fate fall on thee.

Lo here the iron winter of curst caste
    Has made men into things that creep;
The leprous beggars totter trembling past;
    The baser sultans sleep.

Not for a thousand years has Freedom's cry
    The stillness of this horror cleaved,
But as of old the hopeless millions die,
    That yet have never lived.

Man has no leisure but to snatch and eat,
    Who should have been a god on earth;
The lean ones cry; the fat ones curse and beat;
    And wealth but weakens worth.

O Heaven, shall men rebelling never take
    From Fate what she denies, his bliss?
Cannot the mind that made the engine make
    A nobler life than this?[29]

In this poem, Ross questions whether the negative characteristics that he observes in the Indian people are due to the collapse of a once-proud civilisation. He ponders whether the same thing will happen to the 'vigorous populations' of Europe, or whether science can find a way to prevent this. Ultimately, Ross is expressing in poetic form his theory that 'men can better society and themselves and banish most of the ills which now afflict them by precisely the same scientific methods as they now employ in the making of their wonderful engines'.[30] Such a philosophy was in line with Ross's appreciation of his uncle, Captain Barwell, that is, that 'nervous tricks' and 'other weaknesses' of the human character could, and should, be exorcised though science. While Ross's attitude towards the country of his birth is both abhorrent and ignorant, it was not uncommon for people at that time and with his background to express similar sentiments.[31] In writing 'India', Ross also demonstrated his ability to use poetry both to extol and communicate his scientific ideas, however ill-conceived.

It was around this time (1881–82) that Ross also wrote the complementary poems 'Thought' and 'Science', exploring some of the notions that he believed poetry and science, and the disciples of these disciplines, to possess.[32] In 'Thought' he writes:

> Thou wert not nurtured 'mid the marish[33] flowers,
> Or where the nightshades bloom, or lilies blow:
> But on the mountains. From these keeps of snow
> Thou seest the heavens, and earth, and marts and towers
> Of teeming man; the battle smoke that lours[34]

While in 'Science':

> And, with no lying lanthorne's[35] antic[36] glow,
> Reveal the open way that we must go.[37]

From these two poems, it is clear that even at this stage in his career Ross believed science and poetry to offer complementary worldviews. From its lofty position, poetry (represented here by 'Thought') is able to reveal the many different paths that are open to us, while science provides the light to determine which of these paths should be taken.

At around this time, Ross also rediscovered his interest in mathematics, following the chance reading of *The Orbs of Heaven*, a book that he had won as a prize while studying at Springhill.[38] Having rekindled his interest in the subject he once more ventured to Higginbotham's bookshop, this time to purchase all of the mathematics books that he could find. In studying these books, he found that the subject came quite naturally to him; however, rather than acting as a source of inspiration, this discovery merely frustrated Ross, who at the age of 25 felt as though he had perhaps missed his true calling. As such he now considered mathematics to be 'quite useless to me, either as a medical man or as a writer'.[39] Setting aside mathematics for the time being, Ross once more focused his ample leisure time on the pursuit of poetry. This included a continuation of his epic poem 'Edgar', begun at 18, which in 1883 he published, at his own expense, alongside another piece as *Edgar, or the New Pygmalion; and the Judgement of Tithonus* via Higginbotham's bookshop, which was no doubt ever grateful to the young medic for his continued patronage.[40] The publication, however, received no reviews, a very limited print run, and an even smaller readership.

By 1888 Ross had been stationed in India for six years, yet due to an administrative error there were no junior doctors to cover any extended leave of absence. Denied the furlough to which he was entitled, Ross now found himself suffering from a great despondency, regretful that he was yet to achieve anything of significance through either his medical practice or his writing. The following lines from his later collection of poetry, *In Exile*, were written about this period, and reflect his feelings that he was 'living in a cemetery':[41]

This is the land of Death;
    The sun his taper is,
Wherewith he numbereth
    The dead bones that are his.[42]

After threatening to resign from the Indian Medical Service, he was granted a year's immediate furlough, returning to England on 28 June 1888. Following his despondency, Ross was now more determined than ever to achieve something prodigious in the field of medicine. Upon his return to London he therefore enrolled at the Royal College of Surgeons and Physicians, scraping a pass in their newly established Diploma in Public Health. During December 1888, Ross also met his future wife, Rosa Bessie Bloxam, and following a rapid courtship they were married in London on 25 April 1889.[43] After honeymooning in Scotland, Ross applied for an additional two months of leave so that he could study bacteriology under Emmanuel Edward Klein, a scientist who many now view as the 'father of British microbiology'.[44] In the late summer of 1889 Ross headed back to India, equipped with the scientific skills that he believed would help him make his mark on the world.

During his time in England, Ross succeeded in publishing his romance novel *The Child of Ocean*, following a somewhat farcical agreement with a small publishing company.[45] He signed the contract on the day before he departed, meaning that he was still making edits to the draft on the voyage back to India. While the first edition of the book sold out, the agreement that Ross had hastily signed meant that he was liable for the costs of both advertising and the contractually undefined 'failure' of the book. Despite this charade, the book was at least seen by the press, where it received generally favourable reviews; however, the inconsistency of their tone annoyed Ross, who felt aggrieved by the inability of readers to fully understand his intent. This incapacity to communicate his intentions is something that would befall Ross on numerous occasions, not least with the 1911 publication of *Lyra Modulata*, a collection of ten poems written in a phonetic spelling of his own invention.[46] As a result, several of Ross's poetic works were republished in different volumes, normally at the request of a confused readership who wrote either to Ross or to his publishers requiring further context.

Upon arrival in Madras on 6 September 1889, Ross was immediately ordered to report to Burma[47] for field services, leaving behind his wife, who had never before travelled away from home. Eventually she was allowed to join him in Pakoko,[48] but Ross was quickly sent off on location again, once more leaving his wife behind, while he escorted and treated convoys of causalities on marches through the Burmese countryside. Eventually Ross was given a semi-permanent position at Bangalore,[49] as the medical officer to the local general and his staff. During this time Ross was once again writing prodigiously, including a drama called *The Deformed Transformed*, which was a continuation of and homage to Byron's unfinished final play of the same name.[50] Ross paid for fifty copies of the play to be printed by the *Bangalore Spectator* press, shipping some copies home to his family, whereupon his brother-in-law set about finding a publisher for it. It was eventually printed in a small run and received mixed reviews.

Ross had also begun work on *In Exile*, a poetic travelogue of his time in India, for which he devised a new rhyme scheme which he called the 'sonnetelle', which he later described in his autobiography:

I settled upon the three-foot lines in three quatrains, which was finally adopted (really four crotchet-feet to the bar-line). This scheme contains only thirty-six feet in each 'sonnetelle,' and therefore seems to be capable of the utmost fire of concentration possible

– and in my opinion concentration is perhaps the greatest virtue in verse [...] The partition of each sonnetelle into three stanzas allows of a beginning, a middle, and an end – agreeable to my love of form [...] The whole was designed to contain descriptions of scenes and people, thoughts on science, art, and affairs, satires, and paeans, and to end in an integration of opinion which was to be the final philosophy![51]

As well as describing the form of the sonnetelle, this passage reveals what Ross hoped to achieve with *In Exile*; in addition to being a travelogue, he believed it would enable him to further develop his philosophies and present them to his public (not unlike Davy's *Consolations in Travel*, which was written towards the end of his life). In reading Ross's words, it is clear that he has a deep understanding of poetic form, and that he also considers the musical lilt that the construction of his poetry will produce. While Ross uses a variety of forms and metres across his other poetry, throughout *In Exile* he sticks to the strict rhythmical pattern of the sonnetelle, as illustrated by the following verses, taken from the 'Death Stanzas':

This moonèd Desert round,
    Those deeps before me spread,
I sought for hope and found
    Him beautiful, but dead.

In this resounding Waste,
    I sought for Hope, and cried,
'Where art thou, Hope?' – Aghast
    I found that he had died.

I cried for Hope. The Briars
    Pointed the way he'd gone;
Cold were the Heav'nly Fires,
    Colder the numb-lipped Moon.[52]

Ross wrote these lines at the beginning of 1893, and as can be seen, at the age of 38 he had once more fallen into despondency with regard to his perceived lack of achievements, noting that 'I was again not at all well then. I had tasted of great sciences and arts, and had myself attempted something in some of them, but everything I had tried had failed.'[53] However, in 1894 he would return to England once more on furlough, setting in motion a series of events that would bring him the success that he so desired.

O million-murdering death

Malaria is an intermittent fever, symptomised by chills, shivering, vomiting, and a high temperature. It is an extremely dangerous disease that can be fatal if not treated quickly, leading to serious complications including severe anaemia, through the destruction of red blood cells; cerebral malaria, caused by blockages of the small blood vessels to the brain; and respiratory diseases, through damage to the lungs. If not treated properly, those who have been infected can suffer a recurrence some months, or even years, later. Malaria is now known to be caused by the *Plasmodium* parasite; there are many different types of this parasite, but only five cause malaria in humans: *Plasmodium falciparum*,[54] *Plasmodium vivax*, *Plasmodium ovale*, *Plasmodium malariae*, and *Plasmodium knowlesi*. The *Plasmodium* parasite is spread by the female *Anopheles* mosquito;[55] once a person is bitten by an infected mosquito, the parasite passes into the bloodstream, where it makes its way towards the liver. Upon entering the liver it breeds and multiplies, producing hundreds of thousands of harmful organisms called merozoites, which then leave the liver and re-enter the bloodstream, where they infect red blood cells. At regular intervals (usually every 48–72 hours) the merozoites burst out of the red blood cells, releasing more parasites into the blood, and resulting in the chills and intermittent fever that characterise malaria. If an uninfected *Anopheles* mosquito bites an infected human then the malaria is also passed on to the mosquito and the cycle begins again. Many of the foundations of this understanding were laid by Ross, who in turn was indebted to the relationship that he would form with the Scottish physician Patrick Manson when he returned to England in 1894.

As a young child in India, Ross had been exposed to the effects of malaria, with his father contracting the disease and suffering several relapses, including one while engaged in battle.[56] As an adult Ross first encountered malaria when he was a medical student at St Bartholomew's, where he was tasked with conducting the examination of an English woman who had caught the disease in Essex. From the fifteenth to the nineteenth century malaria was present in England, although the draining of marshlands and an increase in the cattle population (which presented an alternative blood meal for mosquitoes) resulted in its eventual disappearance. The patient whom Ross examined was probably one of the last victims of malaria in England, and an excitable Ross caused her to flee from the surgery, such was the intensity of his questioning. When Ross later returned to India with the Indian Medical Services, a large part of the work that he did in Madras was in treating soldiers who were ill with malaria, and while the quinine that was administered was quite successful (at least in the short term), many soldiers died because they failed to get adequate treatment in time.

The word malaria originates from medieval Italian and translates literally as 'bad air'; when Ross began working in India the most common explanation for the transmission of the disease was the inhalation of malignant air, especially that found near hot and humid marshlands. Ross doubted that this was the case, and was instead convinced that malaria arose from a form of poisoning in the bowel, publishing his first paper on the subject in 1893.[57] This research was in direct contradiction to the work that was being carried out by the French physician Charles Louis Alphonse Laveran, who had discovered that malaria was caused by a parasitic protozoan[58] while working in a hospital in Constantine, Algeria.[59] When Ross became aware of Laveran's work some time in 1892, he was initially dubious of the results, failing to identify the guilty protozoan in any of the blood smears from his own infected patients. It is clear that Ross was giving a lot of thought to malaria at this time, and as well as publishing his first medical articles on the subject he also wrote the following verse some time during his appointment in

Bangalore from 1890–93, which was later published in *Philosophies* as 'Indian Fevers':

In this, O Nature, yield I pray to me.
I pace and pace, and think and think, and take
The fever'd hands, and note down all I see,
That some dim distant light may haply break.

The painful faces ask, can we not cure?
We answer, No, not yet; we seek the laws.
O God, reveal thro' all this thing obscure
The unseen, small, but million-murdering cause.[60]

Ross's medical publications on malaria, despite later being shown to be incorrect, had at least revealed his interest in the subject. After returning to England on leave in 1894 he discussed the latest research with his colleagues, and they convinced him that *Plasmodium* did in fact exist, before encouraging him to seek out Patrick Manson, one of the leading authorities on tropical diseases at that time. Manson demonstrated to Ross that Laveran's theory was correct, and that he had not observed any of the parasites in the Indian blood smears because of poor equipment.[61] Crucially, Manson also shared with Ross his hypothesis that the mosquito was somehow involved in the spread of the *Plasmodium* parasite.

Ross had first become interested in mosquitoes around 1883 during one of his first stints in Bangalore, when he found that he was able to rid an infestation in his home by simply tipping over a tub of water in which the mosquitoes were breeding. However, when he told the adjutant of his desire to banish the mosquitoes from the Mess Hall by removing any stagnant water (e.g. flower vases, garden tubs, etc.), he was prevented on the grounds that he might upset nature's balance. Ross's next scientific encounters with mosquitoes occurred in 1890, during his numerous military marches around Burma. Due to his inability to conduct any bacteriological work, he had begun to study mosquitoes seriously, learning to breed them from larvae in a similar fashion to the butterflies that he had raised from caterpillars among the greenhouses of Springhill.

The years 1894–95 marked a critical moment in Ross's life, as for arguably the first time he was being asked to choose between 'Science' and 'Thought'. In addition to his recent publications on malaria, he had just rewritten *The Deformed Transformed* as a romance novel entitled *The Revels of Orsera*, and was also seeking publication for *The Spirit of Storm*.[62] Following the literary success of Rudyard Kipling's short stories, Ross had also begun to experiment in this form, drafting several unpublished efforts including 'The Vivisector Vivisected', a macabre story of a ruthless researcher's attempts to keep a patient alive by pumping him full of hot blood. On his return to India in 1895, Ross ultimately decided to pursue his research into malaria, a decision that was probably influenced by three factors: his conversations with Manson, his unfortunate experiences with publishers,[63] and his receipt of the Parkes Memorial Gold Medal, which he was awarded in March 1895 for an essay on malaria that explored Laveran's theory on parasites and Manson's hypothesis about mosquitoes.[64] Of these three factors, it was arguably the Parkes Memorial Gold Medal which proved the most decisive, providing as it did a concrete example of the external recognition that he so desired.

On returning to India in 1895, Ross found himself posted to medical stations where there were relatively few cases of malaria, meaning that his research initially made slow progress. Gradually, however, he started to make headway, seeking guidance from Manson, with whom he kept up a consistent correspondence, averaging a letter almost every fortnight. In these letters Ross reported his findings, detailing his improved skill in both hatching mosquito larvae and dissecting the insects at various stages of their development, his successful searches for the *Plasmodium* parasite, and his attempts to trace its movements within the mosquito. This extract from a letter sent to Ross on 27 November 1896 demonstrates the extent to which Manson was providing both technical expertise and emotional support:

I would therefore recommend you to try two lines of experiment. First endeavouring to feed or rather provide a house for the living flagellum;[65] second, having found this, to prepare slides of a permanent character showing the flagellum in his cell. Don't forget to search the ova of the mosquito.

I think that the addition of a little breath vapour to your mosquito blood slides will help you very much; try it.

[…] Take care of your health when you go to that village which I hope you will immortalise.[66]

As well as providing scientific support and assistance to Ross via this correspondence, Manson also worked tirelessly to try and secure research leave for him, and was instrumental in securing publication for Ross's findings in the prestigious *British Medical Journal*. In August 1897 Ross began experimenting with a brown and dapple-winged mosquito (i.e. an *Anopheles* mosquito), hatching the larvae and letting these mosquitoes feed on an infected patient, before dissecting them at various stages in their development. On 20 August 1897[67] he discovered some black, pigmented cells in the mosquito's stomach, which by the next day had grown substantially, indicating that the *Plasmodium* parasite was now present in the mosquito. Effectively, Ross had proven that malaria could be transmitted from an infected human to a mosquito, and in celebration of this fact he penned the following sonnetelle:

This day relenting God
    Hath placed within my hand
A wondrous thing; and God
    Be praised. At His command,

Seeking His secret deeds
    With tears and toiling breath,
I find thy cunning seeds,
    O million-murdering Death.

I know this little thing
    A myriad men will save.

O Death, where is thy sting?
Thy victory, O Grave?[68]

Despite drafting this poem almost immediately after his discovery, it took Ross another couple of days to find the words to describe the events to Manson. Ross's instinctive use of poetry to celebrate and contextualise his achievements is in line with his use of the sonnetelle to capture 'thoughts on science', and is also perhaps an apologetic gesture to the neglected path of 'Thought'; an olive branch to help immortalise the accomplishments of 'Science'.

Despite his successes, the cycle of malaria transmission remained incomplete, as it still had to be demonstrated that infected mosquitoes could transmit the disease to humans. As Ross once again found himself posted to locations with a smaller number of malaria outbreaks, he decided to take matters into his own hands, moving his research to birds. Like humans, birds are also susceptible to malaria, via a parasite called *Plasmodium relictum*. By July 1898 Ross was able to prove that the *Plasmodium relictum* parasite was transmitted from infected mosquitoes to birds and vice versa, thereby completing the malaria cycle and formalising the proof that mosquitoes were responsible for the transmission of malaria. Once more Manson was instrumental in helping Ross get this research published, with Ross acknowledging his contributions in the concluding sentence of his 1898 paper 'Report on the Cultivation of Proteosoma Labbé, in Grey Mosquitos':

These observations prove the mosquito theory of malaria as expounded by Dr. Patrick Manson: and in conclusion, I should add that I have constantly received the benefit of his advice during the enquiry. His brilliant induction so accurately indicated the true line of research that it has been my part merely to follow its direction.[69]

A further letter, written by Ross to Manson on 21 March 1898, which begins with the following sentence, corroborates this gratitude: 'My one wish is that you were here to share with me the pleasure which I have experienced yesterday and today in seeing your induction being verified step by step.'[70]

In 1898 Ross threatened to resign from the Indian Medical Service, after once more being ordered to redeploy to another malaria-free environment. However, following a further intervention by Manson, he was instead invited by the Indian government to travel to Assam in north-east India to study an outbreak of kala-azar.[71] Ross hypothesised that mosquitoes were the vectors for this disease,[72] but his experiments were unsuccessful, and it was later shown that kala-azar is instead transmitted by sandflies. By this time the Royal Society had set up a special malaria committee, sending out the English physician Charles Wilberforce Daniels to both confirm Ross's investigations and continue his work. After handing over the reins of his investigations to Daniels, Ross retired from the Indian Medical Service, returning to England with his family in February 1899.

Over a century later, and despite the optimism of his poem, we are still to realise the defeat of malaria that Ross assumed his research had assured. According to the most recent figures from the World Health Organization there are an estimated 214 million cases of malaria worldwide each year, resulting in an estimated 438,000 deaths.[73] While this number represents a 48 per cent decrease in global deaths since 2000, there is clearly still a lot of work to be done, especially in Africa, which accounts for 90 per cent of all malaria deaths, and where more than a hundred children die from the disease every hour.[74]

A bitter Nobel

On returning to England in 1899, Ross took up residence in Liverpool, where he was appointed Lecturer in the newly established Liverpool School of Tropical Medicine. Now that the relationship between malaria and mosquitoes had been categorically proven, much work was being done to try and eradicate the disease, including the extensive drainage of swamps and the use of prophylactic quinine as a preventative medicine. Following his work with the *Anopheles* mosquito, Ross was convinced that it was a fragile creature that could be easily controlled, and he set about devising several strategies to do this. These included a sophisticated mathematical model, which calculated the extermination rate needed to reduce the malaria risk in any designated area. Ross went on to publish these findings in his 1910 book *The Prevention of Malaria*, but at the time his arguments either fell on deaf ears or else went over the heads of his fellow malariologists.[75] Ross would go on to publish several mathematical works in the early 1900s, including *The Algebra of Space*, which he sent to James Clerk Maxwell's great friend P. G. Tait shortly before his death.[76] Ultimately this work would help to form the mathematical foundation for epidemiology;[77] however, at the time his suggestions regarding mosquito control were treated with a similar enthusiasm to that expressed by the Mess Hall adjutant of Ross's early career. As with this earlier incident, Ross did not take kindly to being ignored, resulting in a souring of relations with other malariologists during unsuccessful research trips to Sierra Leone, Nigeria, and India.

Despite his lack of success in these ventures, this was to be an incredibly fertile period in Ross's life in terms of personal accolades. In 1901, and with Manson as his chief sponsor, Ross was made a Fellow of the Royal Society. In the same year he was also elected a Fellow of the Royal College of Surgeons, and in 1902 he was appointed a Companion of the Most Honourable Order of the Bath, receiving his full knighthood in 1911. In 1902 Ross was also promoted to Professor at the Liverpool School of Tropical Medicine, and he was awarded the 1902 Nobel Prize in Physiology or Medicine, becoming the first British recipient. The citation read, in part: 'For his work on malaria, by which he has shown how it enters the organism and thereby has laid the foundation for successful research on this disease and methods of combating it.'[78] However, in being the sole recipient of the award, Ross was to spark a controversy that continues to this day.

In 1890 the Italian physician and zoologist Giovanni Battista Grassi had also begun working on the malaria problem, following his groundbreaking work on determining the life cycle of the roundworm.[79] In 1897, independently of Ross, he established the developmental stages of the *Plasmodium* parasite in the *Anopheles* mosquito, becoming the first person to prove that it was only the female of this species that could transmit malaria. In 1902, when the Nobel committee met to determine the recipients of their awards, they originally decided that Ross and Grassi should share the prize; however, by this time Ross had been campaigning against Grassi's claims for a number of years, writing in a 1900 edition of the Italian periodical *Il Policlinico* that 'Whatever his achievements in other branches of science may be, his claim in regard to malaria may be defined as being those of an energetic, dexterous, and unscrupulous writer who has discovered the discovery of another man.'[80] Given these claims, the Nobel committee appointed an independent adjudicator to investigate, with the German physician and microbiologist Robert Koch coming down on the side of Ross. Today it is widely believed that this decision was incorrect, and that Grassi and Ross should have shared the prize. Yet despite his victory, Ross resented Grassi for the rest of his life. Visiting Ross in 1932 (a few months before he died), the American entomologist Leland Ossian Howard recalls how

> [h]e showed us a big cabinet in which he had systematically filed and indexed all of the papers relating to his malaria work. He swore about the Italians, spoke of Grassi as a damned liar ... He gave us each a copy of his latest paper on the Grassi claims ... Grassi was a damned pirate.[81]

Given Ross's egotism and defensive nature, it is perhaps not surprising that he was unwilling to share credit with Grassi.[82] However, arguably one of Ross's biggest failings was his refusal to give Manson the recognition that he rightly deserved. The support provided by Manson's correspondence, not to mention the fact that the mosquito hypothesis was originally his, made him strongly eligible for such credit. While Ross acknowledged some of this support in his Nobel Lecture, he largely used it as an opportunity

112

to develop his attack on Grassi, castigating his efforts as 'obviously hasty and unreliable' and remarking that they 'exercised no influence whatever in the completion of my own labours'. Even the preamble to the references cited in this lecture reads like a haughty riposte, rather than the gracious acknowledgements of a magnanimous Nobel Laureate: 'This list of works, chronologically arranged, includes chiefly my own writings, many of which are omitted in bibliographies, and such others as are referred to in the text.'[83]

Following the Nobel prize, Ross and Manson's relationship deteriorated, although in truth things had not been the same between them since Grassi had dedicated his 1900 book *Studi di uno Zoologo Sulla Malaria* (*Studies of a Zoologist on Malaria*) to Manson.[84] In this book Grassi described his own version of the history of malaria research, downplaying Ross's achievements and making claims for the importance of his own research. In the dedication, he describes Manson as 'geniale iniziatore delle attuali ricerche', the 'ingenious initiator of the present research'. Ross was outraged by this claim and demanded that Manson make a full refutation; for his part Manson claimed that he had agreed to the dedication in good faith, before he was aware of the contents of the book. In subsequent editions Grassi removed the dedication to Manson, but instead capitalised the following passage in the introduction:

> Spetta però lo si noti, a Manson il grandissimo merito di avere suggerito a Ross di rintracciare il parassita malarico e seguirlo dentro il corpo della zanzara.[85]
> [It is worth noting, in Manson, the great merit of suggesting that Ross trace the malarial parasite and follow it inside the body of the mosquito.]

With personal accolades to collect, mathematical models to develop, and private feuds to fight, Ross had scant time for poetry during this period, acknowledging himself that 'ever since 1897 I had had little time for literature'.[86] However, in 1904 he began editing the sonnetelles for *In Exile*, which was first published privately in 1906. Between 1907 and 1913 he would publish a further five collections of poetry, including the first editions of *Fables* (1907) and *Lyra Modulata* (1911). He continued to lecture in Liverpool until 1916, combining this role with that of consultant physician to King's College Hospital, which he had accepted in 1912. Ross was also appointed vice president of the Royal Society from 1911 to 1913, winning the Society's Royal Gold Medal for his achievements in 1909. In 1916 he was also invited to become the vice president of the Poetry Society, marking the occasion by organising a reading of some of his own works, before later being elected to the presidency. However, following a dispute he resigned as president at the end of 1918 without making any noticeable impression or impact in the role.

The First World War was to have a significant impact on Ross, in both his professional and personal life. His first-born son, Ronald Campbell, was killed at the Battle of Le Cateau on 26 August 1914, and Ross's poem 'Father' was written about this loss:

Come with me then, my son;
   Thine eyes are wide for truth:
And I will give thee memories,
   And thou shalt give me youth.

The lake laps in silver,
   The streamlet leaps her length:
And I will give thee wisdom,
   And thou shalt give me strength.

The mist is on the moorland,
   The rain roughs the reed:
And I will give thee patience,
   And thou shalt give me speed.

When lightnings lash the skyline
   Then thou shalt learn thy part:
And when the heav'ns are direst,
   For thee to give me heart.

114

Forthrightness I will teach thee;
   The vision and the scope;
To hold the hand of honour:—
   And thou shalt give me hope;

And when the heav'ns are deepest
   And stars most bright above;
May God then teach thee duty;
   And thou shalt teach me love.[87]

This heartfelt poem brings to mind two of Ross's earlier works. First, the poem's rawness is reminiscent of the sonnetelle that was written following the great malaria breakthrough; in both instances, Ross instinctively turns to poetry in an attempt to make sense of the situation. Secondly, the tone and narrative are reminiscent of 'The Night Ride', both poems painting an emotional portrait of a father and his doomed son, the nameless pursuer here replaced by the faceless spectre of war.

In 1916 Ross left his post in Liverpool and was appointed by the British government as the consultant in malaria to the War Office. While travelling to Thessaloniki in 1917, his vessel was torpedoed by a German submarine just off the Greek island of Ithaca. Ross described this event at a presentation that he later gave to the Royal Institution:

> We were torpedoed off Ithaca at 8 a.m., but escaped on our escorting torpedo boats. We then proceeded to attack the submarines with depth-charges because they could not escape in the land-locked bay. We brought up one by a depth-charge and then sunk her with shells, and rescued 18 of her crew. Two aeroplanes participated in the hunt, and we think we destroyed another submarine by gunfire. Our ship, the Chateau Renault, was sunk.[88]

Two days after this incident Ross decided to travel to Delphi in mainland Greece, in order to visit the Temple of Delphi, located on the south-western slope of Mount Parnassus. The Temple had been the home of the Pythia, the high priestesses of Ancient

Greece, who had channelled the voice of the god Apollo. After travelling to the site, Ross asked Apollo what the purpose of war was. As well as being the god of poetry, Apollo was also the god of giving the science of medicine to man, and so evidently his opinion was of great interest to Ross. His (alleged) reply was captured by Ross in poetic form:

> But man would not be taught
> And, climbing higher, fell —
> A fancied heaven sought
> But reach'd a real hell[89]

This is a short extract from a longer poem about the whole torpedo incident and Ross's subsequent epiphany, although in truth it is quite a laboured effort that suffers from the lack of concentration that Ross himself found to be so fundamental to effective verse. It did, however, inspire Ross's friend, Arthur Conan Doyle, to write a poem of his own, in a somewhat less pompous fashion:

> I've read of many poets, Latin, Greek,
>   And bards of Tarragona or Toledo,
> But you, dear Ross, are surely quite unique,
>   Blown to Parnassus by a Boche torpedo.[90]

Myths and fables

After the war Ross continued to work towards assuring his place in the annals of scientific and medical history. As previously noted, his vendetta with Grassi continued until the end of his life, and he also worked hard to try and establish his mathematical theories and models, neither of which would be fully appreciated during his lifetime. Despite his reputation (or perhaps because of it), he never managed to secure the well-paid position that he believed he was entitled to. When the Ross Institute was opened in Putney on 15 July 1926 by the Prince of Wales, Ross was given the post of director-in-chief, a position that he retained for the remainder of his life,[91] but which did not pay particularly well – he eventually

116

had to sell his own medical papers to support his wife and family.[92] In 1923 Ross published his autobiography, *Memoirs: With a Full Account of the Great Malaria Problem and its Solution*, which despite providing fascinating source material for much of his early life, does not perhaps have the most reliable author, as the immodest title might suggest.[93] In addition to his own memoirs, Ross later subsidised Mégroz's favourable biography, *Ronald Ross, Discoverer and Creator*, and also continued to write poetry, composing 'My 70th Birthday' on 13 May 1927:

> Laburnum, lilac, chestnut, may
> Round my window welcome day,
> Golden, purple, white and red
> Each with pearl drops dowered.
> Blazing leaps our lord the Sun.
> They, like me, have just begun –
> Who is that old dotard Time
> Dares to rob us of our prime?[94]

Shortly after writing this poem Ross suffered a stroke, following which he was confined to a wheelchair. However, despite his ill health he continued to publish several scientific and literary works, including *Poems* in 1928, which featured a collection of poetry written over the preceding fifty years. His wife died in 1931, and Ross followed soon after, dying on 16 September 1932 at the Ross Institute. He is now buried beside his wife in Putney Vale cemetery.

Almost a year after his death, the Poet Laureate John Masefield, whom Ross had befriended during his time with the Poetry Society where he assisted Ross with the proofs of *Philosophies*, organised a commemorative service at the Church of St Martin-in-the-Fields, London.[95] Twenty-five years later Masefield organised a luncheon at the Ross Institute of Tropical Hygiene on 13 May 1957 to celebrate the centenary of Ross's birth, where he read the following poem that he had composed in tribute to his late friend:

> A century since, this hater of mis-rule,
> This strive after wisdom, first began

117

His battle with the miseries of Man.
Diseases and the doings of the fool.

To him, to toil, in Fortune's fell despite,
Alone, at bay, in jungles of despair,
Illumination laid a secret bare,
The stars of Heaven sang him into light.[96]

Ross is not an easy man to categorise, and his behaviour and achievements were often somewhat contradictory. Despite his medical training, he considered himself to be a scientist rather than a clinician, with large parts of his life, especially after his Nobel prize success, dedicated to fighting petty feuds and bemoaning a perceived lack of recognition. This lack of appreciation, especially in terms of remuneration, often left him bitter and mean-spirited. He never achieved the financial security that other famous scientists of his generation accomplished, and he seemingly held such successes against those who did. For example, Ross complained bitterly about the £30,000[97] in research grants that the English physician and scientist Edward Jenner had received from the British government for his development of the smallpox vaccine, an achievement that Ross thought less impressive than his own accomplishments. Such envy also extended to Manson; despite the great altruism that had been shown to him by his former mentor, Ross resented the fact that Manson's medical practice (and finances) had thrived while his had not. Sadly, towards the end of his life Ross cemented these ungenerous feelings by writing the *Memories of Sir Patrick Manson*, which he used to further downplay Manson's role in the malaria research.[98]

From his letters and his autobiography, as well as from the accounts of others, Ross comes across as egotistical and defensive, yet from his writings it is clear that he could be jovial and even self-deprecating. This complex personality revealed itself in both his scientific and literary writings, which ranged from groundbreaking medical insight and poignant reflections on the nature of loss, to petty-minded score settling and pretentious experiments with phonetics. What is clear from Ross's writings is that he viewed science and poetry (or 'Science' and 'Thought') as two sides of the same coin, and that for him one could simply not exist without the other. In both his scientific and his literary writings he was influenced and supported by others, either through the teachings of Manson or his reading of Goethe and Byron, although he was often reluctant to acknowledge these influences. Throughout his career Ross used poetry as a means through which to communicate his scientific research and his wider philosophical beliefs, both to himself and others, and as Ross himself once observed in an address to the Royal Institution:

To sum up, then, Science and Poetry dwell together. We shall reach Truth by seeking Beauty, and Beauty by seeking Truth. Nor shall we attain one without the other.[99]

5

The reluctant poet: Miroslav Holub

lost in the landscape
where only surgeons
write poems.

From 'Vanishing Lung Syndrome' by Miroslav Holub[1]

Ancestral heirlooms

Miroslav Holub was born in Pilsen, Czechoslovakia,[2] on 13 September 1923. His father was a lawyer for the Pilsen Directorate of State Railways and his mother was a high school teacher who taught French and German. There is not a great deal written about Holub's childhood and early adulthood, and as a rule he did not like to talk about his personal life. However, in 1994 he published *Ono se letělo*, a collection of poems inspired by his hometown of Pilsen; compiled to mark the 700th birthday of the city, these poems, which were written at various stages in Holub's life, are interspersed with recollections of his childhood and of growing up in the city. Translated into English by Ewald Osers and published in 1996 as *Supposed to Fly: A Sequence from Pilsen*, this collection offers a fascinating insight into both Holub's early life and his poetic sensibilities, and along with interviews conducted in his later life, it is one of the main sources for this chapter.[3] The title of *Supposed to Fly* comes from a story in the book about a racing pigeon that goes missing from a race because it didn't realise it was supposed to fly, but it is also a nod to the surname of the author, which means pigeon in Czech, an identity that Holub would continue to play with in his poetry and other literary writings, for example his 1977 English-language poetry collection *Notes of a Clay Pigeon*.[4]

Holub appears to have had a relatively comfortable childhood, describing this time as being 'Neither hard nor sentimental, but cotton-wool-wrapped and equipped with a garden, and with two lavatories, albeit unheated ones, which delayed my way in the world.'[5] The humour that Holub displays in this quotation is typical of the poetical style that he would come to develop, but it also highlights a fundamental belief of his philosophy, that is, that difficult personal situations resulted in better poetry. As he later stated in an interview with the South African journal *New Coin Poetry*:

> When one is living under oppression in more or less difficult sociological or political situations one is usually inclined to write more vigorous and even, which is funny, more optimistic kind of poetry. When you are left alone, when you are free, you relax. You may even become sort of sour, dressed and nihilistic. I have the feeling that poets in the real happy situation lose a part of their essence.[6]

The relative privilege of Holub's youth meant that his family were able to take regular Sunday trips out into the surrounding countryside, while his father's position in the railway guaranteed them higher class tickets. During these trips Holub would occupy himself by exploring the nearby area while daydreaming of being a famous painter or archaeologist. On one such expedition he came across a group of burial mounds from the Hallstatt period, dating from some time between 800 and 500 BC, and as he climbed around the surrounding stones he found it comforting that this historic violence lay buried and silent beneath the ground upon which he now walked. He later observed that 'violent prehistory transforms itself into gentle instructive material suitable, or indeed recommended, for all young boys and girls'.[7] This interest in history, and more importantly in learning from the past while simultaneously realising its transiency, is something that would greatly influence Holub's future writings and philosophies.

Holub believed that our history was not passed down through

ancestral heirlooms, but rather that it was exchanged via our genes, and thus by its very nature was inescapable. Speaking later in life, he stated that

> I believe that the real eternity is in the human genes [...] For me, history is an inevitable and necessary precondition to any understanding [...] To have a sense of history does not automatically make us any stronger, but it does enable us to understand our own weaknesses better – so that we may perhaps be stronger than our ancestors.[8]

These themes are also reflected in his poem 'Great Ancestors', which ends with the following lines, the 'them' and 'they' referring to our human ancestors:

> Anyway,
> we've picked up more genes from viruses
> than from them.
>
> They have no strength.
> And we must be the strength of those
> who have no strength.[9]

As well as his youthful expeditions among ancient burial sites, Holub was also a keen collector of butterflies and rocks, the latter of which he chemically analysed in the family bathtub. He seems to have taken particular pride in his butterfly collection, catching them in a green net or raising them from caterpillars before carefully pinning them on limewood panels, then labelling each of the specimens with tiny tags to identify both their name and the circumstances of their capture. In his poem 'Boy catching butterflies' he writes:

> With a wretched net over his shoulder
> he wanders from tree to tree,
> over equators and poles,
> he hops and skips, and sees the sun,[10]

122

While the nature of his collecting and archiving perhaps hinted at a future scientific career, it is the unbridled joy of his endeavours in pursuing this hobby that is at the forefront of Holub's writing. Sadly, the butterfly collection eventually disappeared, or as Holub recalls:

Unfortunately the world is real. Unfortunately within ten years all my butterflies in their twenty-five boxes were eaten by museum carpet-beetles and other parasites which nurture a grudge against human immortality. Unfortunately all that's left is the pins, the boxes and the labels. And so I started writing poems again. Poems aren't eaten by anything, except stupidity.[11]

As a child and young adult, Holub found his poetic interests strongly influenced by his Francophile mother, especially her appreciation of the French poets Saint-Pol-Roux and Tristan Corbière. Holub was likewise fond of the local poet and composer Jindřich Jindřich, taking part in a primary school performance to mark Jindřich's 55th birthday by singing some of his verses as he watched from a nearby window. As well as poetry and music, Holub was also interested in the theatre, and while studying in Pilsen at the Czechoslovak State Classical Grammar School he would sell tickets for the local auditorium, making use of the complimentary ticket that he received for every ten sold in order to become the self-styled 'theatre expert' of his school.

On 15 March 1939, when Holub was a 15-year-old grammar school student, he found the relative comfort of his unheated outdoor toilets suddenly removed by the Nazi occupation of his home town. This brought with it not only a totalitarian regime and the threat of the Gestapo, but also the risk of death from Allied bombing; the large Škoda works on the outskirts of the city had been commandeered to produce armaments for the German forces, becoming a target of strategic importance. However, due to a series of unsuccessful bombing attempts the Škoda works remained virtually undamaged throughout the Second World War. One such event took place on 16 April 1943, and involved 691 tonnes of bombs being dropped on a large mental hospital that was mistaken for the factory.[12] In the same raid one of Holub's schoolfriends was killed as his home was blown to pieces. Sifting through the rubble, Holub found that his copy of the poem *Manon* by the Czech writer Vítězslav Nezval, which he had recently lent his friend, had somehow remained entirely intact, and he kept the book as a *memento mori* of the tragic events.

The Nazi occupation also restricted the books that Holub had access to, and in addition to the works of Albert Einstein and Franz Kafka (among several others) being entirely removed, many of the permitted textbooks suffered heavy censorship. This included blotting out the acknowledgement of any Jewish scientists, and as Holub notes: 'This inking out resulted in exceedingly strange and inspirational sentence shapes, of the kind "The discovery of and search for new elements was largely due to the outstanding chemist...... (1834–1907)"'[13]

Many Czech nationals took an active stand against the Nazi regime, including a large student rebellion, and on 28 October 1939 (the 21st anniversary of the independence of Czechoslovakia from the Austro-Hungarian Empire) a key moment in the resistance occurred, with demonstrations taking place across Prague. To punish the protestors, the Nazis executed nine of the student leaders and shut down all of the Czechoslovakian universities.[14] In 1942, with options now limited after his graduation from the Grammar School, Holub started work as an assistant in a wood store at Pilsen railway station. During the Allied bombing he would peep out from the air raid shelter to look at the bombs that were being dropped on his home town, capturing this series of events in his poem 'The Bomb', which begins:

Murder in the lithosphere.
Clay burst from the rock,
fire flowed from the clay.[15]

At around this time Holub was also expanding his poetic style, experimenting with the surrealism of Nezval, while likewise being influenced by a variety of French poets, whose works he enjoyed via the translations of the Czech writer and publisher Karel Čapek. On 6 May 1945, while the rest of Czechoslovakia was being liberated by the Soviets, Pilsen was freed from Nazi Germany by the advancing American troops of General George Patton. There was minimal resistance to the Americans' advance, other than from a group of SS snipers perched in the tower on top of the Czechoslovak Hussite church in the centre of the town. During this exchange of fire Holub recalls seeing an American soldier hit in the leg before being carried off in an ambulance to receive treatment. Years later, when travelling in America in the 1960s, Holub met this same soldier while visiting a small cheese farm in rural Wisconsin.

After the war Holub wanted to pursue a profession that would represent the greatest benefit to society, enrolling at the Faculty of Natural Sciences at Charles University in Prague, before switching to the Faculty of Medicine at the same institute in 1946. While at university Holub continued to write poetry, his style still influenced by the surrealist nature of Nezval and the Czechoslovakian avant-garde movement of the 1940s. Some of his first published poems appeared in the journal *Kytice* (*Garland*), which was then edited by the Czech poet and writer Jaroslav Seifert, who would go on to win the 1984 Nobel Prize in Literature. In 1948 Holub came third in a national student poetry competition, but like the chemists of his childhood textbooks he now found himself the victim of censorship. The Czechoslovakian coup of February 1948 resulted in Czechoslovakia being placed under the command of the Communist Party of Czechoslovakia (KSČ), which was allied with the Soviet Union, bringing an end to many civil freedoms, including the prohibition of any literature that was deemed contrary to the now official literary doctrine of Socialist Realism. The surrealist poetry of Holub, and of most of the other Czech students, was now considered inappropriate, and so rather than awarding any prizes, the communist student leader instead dissolved the national students' union. In protest, Holub took a poetic vow of silence and stopped publishing his poetry.

A cultural thaw

During his silent poetical protest against the oppressive nature of the communist regime, Holub devoted himself to science, receiving his Doctorate of Medicine from Charles University in 1953. Holub's interest in communicating science to a wider audience was further expressed in that same year, as he became the executive editor of *Vesmír* (*Universe*), a popular science magazine.[16] *Vesmír* was first published in 1871, and presented current scientific research in a style that was accessible to the general public. After receiving his MD, Holub worked as a pathologist at the Bulovka hospital in Prague, before joining the Institute of Microbiology at the Czechoslovak Academy of Sciences in 1954 (also in Prague), where he began work on his doctoral thesis in the field of immunology. In 1958 he achieved two significant landmarks: he published his PhD and had a paper accepted by the prestigious scientific journal *Nature*, both of which concerned the production of antibodies by lymphocytes.[17]

When Joseph Stalin died in March 1953 he was replaced by Nikita Khrushchev, who introduced his policy of de-Stalinisation, which involved a weakening of the totalitarian regime across the USSR. The so-called 'Khrushchev Thaw' included a cultural component, in which suppressed Soviet writers began to publish again.[18] While the special intensity of Stalinism in the country prior to the events of 1953 meant that this cultural thaw was delayed somewhat in Czechoslovakia, by the end of the 1950s there were many scholars and writers who were trying to liberate their society by strongly voicing their concerns and arguing for greater freedom of speech. It was within this environment that Holub re-emerged as a poet, one who was now influenced by both the Czechoslovakian avantgarde movement and the poetic realism of the French screenwriter and poet Jacques Prévert. As Holub noted in an interview towards the end of his life:

> But when I discovered Jacques Prévert in the 50s, I was more than enchanted: I understood! I realized that poetry may be not

only a state of mind, but also the state of the world reflected in an event. Prévert was for me an assurance that I might be able to write poetry.[19]

Similarly, in an essay published in *Večerní Praha* (*Evening Prague*) in 1963, Holub remarked that

> I prefer to write for people untouched by poetry ... I would like them to read poems in such a matter-of-fact manner as when they are reading the newspaper or go to football matches. I would like people not to regard poetry as something more difficult, more effeminate, or more praiseworthy.[20]

With regard to the surrealist influences on his poetry, he noted that 'I was always ravished by surrealism, but my own surrealism is also a reaction against the reigning absurdity of Marxist ideology, or against the primitive allusions of present-day pseudosciences and spiritual profanities.'[21]

In 1958, the same year that he finished his PhD, Holub also published his first collection of poetry, *Denní služba* (*Day Duty*). One of the poems from this collection, 'Casualty', translated by Ewald Osers, begins:

> They bring us crushed fingers,
> *mend it, doctor.*
> They bring burnt-out eyes,
> hounded owls of hearts,
> they bring a hundred white bodies,
> a hundred red bodies,
> a hundred black bodies,
> *mend it, doctor,*[22]

In this poem, there is nothing esoteric or 'difficult' about the words that Holub uses to capture the raw sentiments of a doctor working in a casualty department. This is poetry at its most matter-of-fact, an insight into the working life of the people; the state of the world reflected in an event.

In addition to the influences of Nezval, Seifert, and Prévert, Holub also took inspiration from the American writer and medical doctor William Carlos Williams,[23] whose prescription-pad poems (so called because they were written between house calls on small prescription pads) served as key models in Holub's poetic development;[24] indeed, the title of his first collection, *Day Duty*, might be taken as an homage to this period of Williams's work. Holub rejected rhyme and metre in favour of a more experimental, surreal, and yet also personal experience. Several of these ideals are expressed in an article that he wrote for the September 1956 edition of *Květen* (*May*), a poetry journal that was set up by Holub and a group of other young Czech poets in search of the 'poetry of the everyday life'.[25] In this article Holub states that 'Only by capturing life around us we may be able to express its dynamicism, the immense developments, rolling on around us and within us.'[26] This style of directly capturing surrounding life was in opposition to the 'official' style of poetry (i.e. Socialist Realism) that was still heavily promoted by Czechoslovakia's communist regime, and while he continued to be inspired by the surrealists, another early twentieth-century movement also had a profound influence on Holub's work. Imagism was an English and American poetry movement that, while short-lived, subsequently had a profound influence on Modernist poetry, with imagist anthologies including work by Ezra Pound, James Joyce, Amy Lowell, and the aforementioned William Carlos Williams.[27] In reaction against the poetry of the immediate past, which was considered to be too formulated around unclear abstractions, the Imagists used common, everyday speech and aimed for succinctness and concrete imagery, and with it the establishment of new rhythms. The influence of Ezra Pound and the other Imagists on Holub is expressed in the following extract, taken from an interview in which he discussed the development of his unique style:

> My first education in literature was Greek, classical Greek. Strict metres bring some limitations to the formation of images and of ideas. I reject those *a priori* limitations, I want to be free in finding out the contrasting situations which form the ruling metaphor of

the poem. On the other hand a poem is definitely not just prose chopped into shorter lines. Even free verse has rules and these rules must not be upset, otherwise you get lost in some sort of inaccessible or less accessible prose.[28]

Holub was not a member of the Communist Party, which made him a second-class citizen in his own country, and as such he was denied the opportunity to travel abroad on research trips and to international scientific conferences until the early 1960s. At this time he was finally granted permission to travel, receiving invitations to work in research laboratories in both Germany and the United States. As his international reputation as a scientist began to blossom during this period, so too did his standing as a poet. Among other appearances, Holub was invited to read his poems at the 1965 Spoleto Festival in Italy, the 1967 Lincoln Center Festival in New York, and the 1968 Harrogate Festival in England. Holub was immensely prolific during this period, producing several collections of poetry in Czech, including *Achilles a želva* (*Achilles and the Tortoise*) in 1960 and *Tak zvané srdce* (*The So-Called Heart*) in 1963. By the late 1960s some of his poetry had also been translated into English, and his first collection, *Selected Poems*, was translated by Ian Milner and published by Penguin in 1967.[29] He also published two books of what he called 'semi-reportage' about extended visits to the United States: *Anděl na kolečkách: poloreportáž z USA* (*Angel on Wheels: Sketches from the U.S.A.*) in 1963 and *Žít v New Yorku* (*To Live in New York*) in 1969.[30] Holub considered travel to be one of his greatest pleasures in life, yet it was a lengthy and laborious process for a Czech national who was not a member of the Communist Party to obtain the exit visas that were required; that Holub travelled so extensively during this period points to both his reputation and also his desire and drive to ensure that travel was possible.[31]

Despite his success as a poet, Holub now considered science to be his vocation, with poetry his pastime. In a 1967 interview for the political and cultural magazine *The New Leader*, Holub revealed that despite the Czech Writers Union offering him a stipend equivalent to his salary as a research scientist (so that he could devote his time instead to poetry), he had turned them down. In the same interview, Holub also revealed that he thought science and poetry to have an 'uneasy relationship', observing that 'I try to hide the fact that I write verse. Scientists tend to be suspicious of poets; they feel that poets are, somehow, irresponsible.'[32] This suspicion would turn out to be well founded, as despite his new-found celebrity, Holub's reputation and livelihood were about to be severely jolted by the events of 20 August 1968. On this date the USSR led Warsaw Pact troops in an invasion of Czechoslovakia to crack down on reformist trends in Prague, ending the delayed 'cultural thaw' under which Holub's scientific and poetic prowess had been allowed to bloom.

Before and after

The Warsaw Pact invasion was in reaction to the Prague Spring, itself set in motion in January 1968, when Antonín Novotný was forced to stand down as the general secretary of the KSČ, a post that he had held since 1953. Novotný was replaced by Alexander Dubček, and in April 1968 his government announced its plans for a new model of socialism – one that would allow freedom of speech and remove state controls over industry. For the next four months there was increased freedom in Czechoslovakia, a period that is now referred to as the Prague Spring. Dubček also stressed that Czechoslovakia would stay in the Warsaw Pact, but at a meeting held in Bratislava on 3 August 1968, the general secretary of the Central Committee of the Communist Party of the USSR, Leonid Ilyich Brezhnev, announced that the USSR would not allow any Eastern European country to reject communism.[33] On 20 August 1968 500,000 Warsaw Pact troops invaded Czechoslovakia. The Czechoslovakians did not fight their invaders; instead they stood in front of the tanks and put flowers in the soldiers' hair.[34] Despite their aversion to violence, 137 Czech civilians were killed and 500 were seriously injured during the invasion, following which Dubček was replaced by Gustáv Husák as leader of the KSČ.

One of the several catalysts that led to the Warsaw Pact invasion of Czechoslovakia was the publication in June 1968 of the 'Two Thousand Words that Belong to Workers, Farmers, Officials, Scientists, Artists, and Everybody', also known as the 'Two Thousand Words' manifesto. Written by the Czech journalist and writer Ludvík Vaculík in the middle of the Prague Spring, this manifesto called for grassroots civic activity and was effectively a criticism of the KSČ, outlining how it had failed the Czech people and the steps that it needed to take in order to improve. The following passage gives a good indication of the overall sentiment of the manifesto:

> Since the beginning of this year we have been experiencing a regenerative process of democratization. It started inside the communist party, that much we must admit, even those communists among us who no longer had hopes that anything good could emerge from that quarter know this. It must also be added, of course, that the process could have started nowhere else. For after twenty years the communists were the only ones able to conduct some sort of political activity. It was only the opposition inside the communist party that had the privilege to voice antagonistic views. The effort and initiative now displayed by democratically minded communists are only then a partial repayment of the debt owed by the entire party to the non-communists whom it had kept down in an unequal position. Accordingly, thanks are due to the communist party, though perhaps it should be granted that the party is making an honest effort at the eleventh hour to save its own honor and the nation's.[35]

The manifesto quickly became the symbol of the Prague Spring movement and was signed by a number of famous Czechoslovakian writers, scholars, and scientists, among them Holub. By signing the 'Two Thousand Words' manifesto, Holub had made a very public statement about the communist regime that was presiding over his country. Following the Warsaw Pact invasion and the installation of Husák as the head of the KSČ, the Prague Spring was officially over, and the freedom of speech that it had brought was no longer in place. Poets, writers, and scholars now once again found their movements and actions restricted and watched, especially those who had spoken so openly against the KSČ by signing the manifesto. Holub now found himself followed and harassed by the StB (the plainclothes secret police force that was controlled by the KSČ); his books were banned, his travel abroad restricted, and in 1970 he was sacked from his post at the Microbiological Institute.[36] In essence he had become a 'non-person', someone whose existence was denied or ignored by the government, as a punishment for disloyalty or dissent. Understandably this was a deeply traumatic time for Holub, and it is the events of 1968 that form the turning point in the title of *Before and After*, the most substantial collection of his poetry to have been translated into English, the first edition of which was published by Bloodaxe Books in 1990.[37]

Comparing the poetry from these two periods, there is a noticeable difference between the poetry of *Before* and that of *After*, especially in the original 1990 edition, which does not include any poetry written after the Velvet Revolution (see page 138 below). The poems of *Before* are more precise and concrete than those of *After*, which as well as adopting more abstract arguments also tend to appeal less directly to both the poetry and the people of everyday life. The poems of *After* are also more subjective and unsure than the objective certainty that is often displayed in the poetry of *Before*; in short, his poetry became less 'scientific' during this period. For example, one of Holub's best-known poems from *Before* is 'The Door', a call-to-arms in which Holub implores the reader to open an imaginary door and see what lies beyond, even if 'there's only the darkness ticking'. This poem is taken from the 1962 collection *Jdi a otevři dveře* (*Go and Open the Door*) and the last lines of the poem, translated into English by Ian Milner, reveal what Holub believed was the worst that could happen if we ever dared to open the door:

At least
there'll be
a draught.[38]

This poem uses very concrete and precise images that appeal to both the poetry and the people of the everyday, a clear and defiant message to his readers that taking a leap of faith might be necessary to welcome in the draught that signals the prevailing wind of change.

The final lines of 'A boy's head', taken from the 1963 collection *Zcela nesoustavná zoologie* (*Totally Unsystematic Zoology*), and translated here by Ian Milner, similarly capture the concrete imagery that Holub used to indicate the possibility of change:

> There is so much promise
> in the circumstance
> that so many people have heads.[39]

Holub's statement in this final stanza is a slightly surreal and ridiculous statement of the obvious, but it is also optimistic; a throwaway line that contains within it a clear message of hope and perseverance to the average Czechoslovakian citizen whom Holub considered to be his reader.

Contrast this with 'Brief Reflection on Death', from the 1982 collection *Naopak* (*On the Contrary*) which, translated by Edward Osers, reads as follows:

> Many people act
> as if they hadn't been born yet. Meanwhile, however,
> William Burroughs, asked by a student
> if he believed in life after death,
> replied:
> – And how do you know you haven't died yet?[40]

The surrealist imagism of Holub's poetry from *Before* is still evident here, but the subtle irony and clear imagery are replaced by a caustic wit and vagueness that characterise many of the poems of *After*. While Holub was a known admirer of William Burroughs and other figures of the Beat Generation, the imagery that he conjures up in this poem is much less concrete than that of the previous examples. There is a level of subjectivity to it, a degree of conflation and complexity that are not associated with the precise scientific approach that Holub adopted in his earlier writings.

Of the 44 poems that make up *Naopak*, 21 have a title that begins with 'A Brief Reflection...' These poems were written between 1971 and 1982, in the immediate aftermath of the fallout from the Warsaw Pact invasion, and the repetition in the titles is an indication that Holub is reflecting on what has gone on *Before*, reassessing his thoughts and 'reflecting' on what he perhaps considers to have been personal misconceptions during this period. While such reflections might be seen as analogous to revisiting a scientific hypothesis, the reflections that Holub embarks on only serve to make less clear those images that had previously been precise and resolute.

This reworking of previous material continues in 1986's *Interferon čili O divadle* (*Interferon, or on Theatre*); written between 1983 and 1986, this collection of poems moves further away from the teachers, the fishermen, and the doctors of his previous poetry and instead focuses on characters such as 'Immanuel Kant', 'Faustus', and 'Mr Punch'. In moving further away from the people and the poetry of the everyday, in terms of both theme and style, these poems perhaps alienated more readers than they connected with. The final series of poems in this collection are presented as short vignettes for the stage; surreal and dark re-imaginings of tales that may be familiar to the reader. 'Fairy Tales' is a particularly bleak retelling of Red Riding Hood, the traditional role of the wolf replaced by 'heads full of teeth', 'shadows of ghosts', and men dressed in black who clear the stage and remove all trace of what had been before. Given his encounters with the StB, this poem might be interpreted as a comment on Holub's position as a non-person, although there remains a vagueness and uncertainty to any message which he is trying to communicate. Further evidence of the self-reflecting nature of this period is given by the final lines of the poem 'The man who wants to be himself' (also from *Interferon čili O divadle*), which translated into English by Osers reads:

I'm probably not worthy of myself, the man mutters.
Perhaps I'd better not admit I'm me.[41]

In 1973 Holub appeared to renounce his previous denunciation
of the communist regime, with a public statement of apology
that apparently confessed his regret at having signed the 'Two
Thousand Words' manifesto, while outlining his respect for
the 'constructive effort of the new party and state leadership'.[42]
This apology ameliorated his status as a non-person and meant
that his literary work could once more be published officially in
Czechoslovakia without censure (although this did not happen
until 1982). While the apology restored some of his 'privileges',
the apparent rejection of his principles was not well received by
his fellow writers and scholars, many of whom felt that he had sold
out his liberal ideals. However, it now seems almost certain that
Holub never wrote such an apology. Peter Bridges, an American
diplomat and friend of Holub, revealed in his memoir *Safirka: An
American Envoy* that Holub considered the apology to be a forgery
by the StB, produced using details provided by Holub's second
wife,[43] whom he eventually realised was a police spy and subsequently left.[44] This was further corroborated by Holub himself in
conversation with the poet and essayist Wallis Wilde-Menozzi and
the academic Jan Čulík, to whom Holub stated that the apology
was a fake and that he had been denied the right to a disclaimer.[45]

In 'The man who wants to be himself', Holub's reluctance to
admit to who he is speaks to a resignation that even if people knew
the truth, there would still be doubters. Earlier lines in this poem
also mention a desire to 'collect the teeth of extinct species' – indicative perhaps of a desire to return to the ancient burial sites of his
youth and to the certainty that accompanied the poetry of *Before*.
The final lines of 'Interferon', again translated into English by
Osers, further evidence Holub's desire to return to something that
he knew he could not:

you're only imagining all this,
look, the butterfly's already

bringing the flowers back ... and
there's no other devil left ... and
the nearer paradise...

He believed, and yet he didn't.[46]

An interferon is any of several related proteins that are produced by the body's cells as a defensive response to viruses, bacteria, and parasites. In titling this collection of poetry (the last to appear in the first edition of *Before and After*), Holub presented a hypothesis that the theatre (and the arts more generally) was a defensive response to the creative oppression that was going on in his homeland. However, in the poems of both *Interferon čili O divadle* and *Naopak*, the clarity that accompanies Holub's earlier poetry is missing, pointing to a lack of certainty and conviction in this belief. In his interview with *New Coin Poetry*, Holub was asked how he saw his poetry in the context of the political upheaval that had taken place across Czechoslovakia, to which he replied that

> [w]hat I noticed in the Communist regime which was in a way a chaotic regime, was disorganisation at every level. The oppression was one thing, but the generalized and legalized mess, moral mess, administrative mess was another. In a condition of general chaos, poetry is the last resource of order. In a completely orderly open democratic society poetry just gets the opposite role, to create some sort of turbulence, some sort of personal chaos which may be interesting.[47]

Holub was considered, and considered himself, to be an activist through his poetry; he believed in the power of poetry to bring about action. In the communist regime he saw disorganisation, and given his opinion that the role of poetry was to oppose the status quo, it might be expected that during this period of censorship Holub's writing would become more ordered and concise. However, as previously stated, the poetry of *After* became less clear and more complex during this period, and while Holub was undoubtedly writing in opposition to the regime, the disorganisation and administrative mess that he was experiencing seems to have influenced and leaked into his poetry. The clear and objective poetry of the everyday that characterised *Before* was replaced by the more reflective and subjective poetry that appeared in *After*.

Talking later in life about his censorship, Holub remarked that

[s]o MUCH depends on the red wheelbarrow, said Wallace Stevens. So much depends on every detail of the personal history. In my experience, the censor of the Communist fifties was close to a hangman; the censor of the mid-sixties was a worried bureaucrat who wanted to live in peace with those 'above', but was easily infected by the ideas from those below. The censor of the seventies was just a mechanism to scare the publishers: an inadmissible literary work, usually the product of an inadmissible author, would simply not be permitted to be distributed, and this might cause the publisher's economic collapse. This was a very efficient kind of censorship which caused many of us to assume the noble status of non-persons.[48]

This statement might make it appear that Holub was more concerned by the censorship of the 1950s than that which followed the Warsaw Pact invasion. However, it also reveals that Holub considered censorship to be somewhat subjective, so while the method by which the censorship of the 1970s was enforced might have been less 'dramatic', it was arguably more effective in silencing Holub as a creative force, certainly in his own country. The first line of this statement refers to an observation by the American Modernist poet Wallace Stevens, who was himself reflecting on the red wheelbarrow of William Carlos Williams's most famous poem, arguing that in order to fully consider its meaning it is important to consider what takes place in the white spaces between the poem's lines.[49] By telling us that so much depends on the red wheelbarrow, Holub is not only reminding the reader of the personal and subjective nature of censorship, he is also instructing us to read between the lines and to look for what remains unsaid in his own commentary. Thus, even though the censorship of the 1970s might have been less 'aggressive' or 'threatening', it was undoubtedly that which

137

affected Holub the most, creating the junction that split the *Before* from the *After*.

Sisyphus and the Velvet Revolution

From the mid-1980s onwards, as general secretary of the Communist Party of the Soviet Union, Mikhail Gorbachev intro-duced two major governmental policies, *Perestroika* (restructur-ing) and *Glasnost* (openness), that resulted in profound changes in economic practice, internal affairs, and international relations throughout the USSR and its satellite states. However, as with Khrushchev's cultural thaw, the KSČ made efforts to ensure that many of these reforms did not take place in Czechoslovakia, and kept tight control over its citizens through the continued suppres-sion of non-conformists. This continued after the fall of the Berlin Wall in 1989 and the subsequent democratic transition of other Soviet Bloc countries such as Poland and Hungary. These events resulted in peaceful protests during November and December 1989, which in turn led to Czechoslovakia ultimately overthrow-ing the communist regime and bringing back democracy for the first time in fifty years, since before the Nazi occupation. The Velvet Revolution, also known as the Gentle Revolution due to the non-violent actions that underpinned it, resulted in the elec-tion of the human rights activist and playwright Václav Havel as president of Czechoslovakia on 29 December 1989, with fully free and democratic elections being held in June 1990. Disagreement between Czech and Slovak politicians led to Czechoslovakia peacefully splitting into two independent countries (albeit without a referendum), with the Czech Republic and the Slovak Republic being established on 1 January 1993.

During this time Holub was still working at the Institute of Clinical and Experimental Medicine in Prague, where he wrote many short essays on various aspects of science and life in general. These included *K principu rolničky* (*The Jingle Bell Principle*) in 1987 and *Shedding Life: Disease, Politics, and Other Human Conditions* in 1997. Following the Velvet Revolution, Holub released three more collections of poetry: *Syndrom mizející plíce* (*Vanishing Lung Syndrome*) in 1990, *Ono se letělo* (*Supposed to Fly*) in 1994, and *The Rampage* in 1997, which was later published in Czech as *Narození Sisyfovo*. *Vanishing Lung Syndrome* was written between 1985 and 1989, before the Velvet Revolution, and so the poems in it are written in a similar vein to the others that were written during the *After* period of Holub's life, while as previously discussed *Ono se letělo* presents a slightly surreal autobiography of the poet.

The poems of *The Rampage* see Holub return to the stripped-back directness of *Before*, and also encompass a re-engagement with poetry both of and for everyday life; 'My mother learns Spanish', the little girl in a red dress who features in 'The end of the week', and the photographs of missing children in the railway stations and bus terminals of 'The journey' are all emblematic of this return to a more direct and objective form of poetry. *The Rampage* also features Holub's honest, cynical, and humorous take on life in Czechoslovakia post-1989. The poem 'At last' is arguably his most direct response to the Velvet Revolution and the events that followed; appearing in the section entitled 'Freedom', it is a surreal and cautious reflection, which ends with the lines:

But we couldn't step out
of our doorways;
someone might cast
a spell on us.

We might even
be hostage
to ourselves.[50]

Given that Holub had already lived through several false dawns with regard to freedom of speech, it is unsurprising that he is restrained in his celebration of the occasion. However, unlike some of the poetry written during the 1970s, this observation is not blunted or obtuse, but rather acute and direct.

The Rampage was Holub's last collection of poetry before he died in Prague on 14 July 1998 at the age of 74, and while it saw him return to the poetry of the everyday, there are still a number of fables and allegories included in it; however, as with 'At last', their message is more direct and immediate than those that characterise the poetry of *After* prior to the Velvet Revolution. The last poem in the book, and hence the last to be published in his lifetime, is 'The birth of Sisyphus', a modern re-imagining of the Greek myth. Sisyphus was the King of Ephyra[51] who was punished in Hades for his various crimes against the gods by having to repeatedly roll a huge stone up a hill, with the stone rolling down to the bottom of the hill as soon as he had brought it to the summit. The fate of Sisyphus is often used as an allegory for pointless and ultimately futile tasks. However, in his well-known essay 'The Myth of Sisyphus', the French philosopher Albert Camus tells us that while Sisyphus is resigned to his fate, we must imagine him as happy, as he has conquered his fate – not by changing it, but by accepting it and yet revolting against it.[52] Holub was an admirer of both Camus and his reflection on Sisyphus, stating as much in an essay that was published posthumously under the title 'Rampage, or science in poetry': 'I share the Sisyphus syndrome as interpreted by the Albert Camus who thought that one must imagine Sisyphus as happy.'[53] In this period of his life Holub gave much thought to the Sisyphus myth, and it is tempting to see elements of Sisyphus in Holub himself[54] – a man who despite constantly being censored and downtrodden by various political systems, continued to push his rock to the top of the hill, and who in so doing revealed the absurdist and humorous futility of life.

The poet or the scientist?

Throughout the various interviews that he gave and essays that he wrote, Holub considered himself to be a scientist first and a poet second. For example, in his 'Rampage, or science in poetry' essay he writes:

> Had I not started both careers in about the same time, with science as the profession and poetry as a 'supporting pastime', I would

be gravitating nowadays towards science anyway, because of its viability and vigour.[55]

This short statement reveals not only that Holub considered poetry to be his pastime, but also that he considered science to be of more value to the wider society, and as such, a 'worthier' pursuit. This is also in keeping with his initial decision to pursue a career in science following the Second World War, as he viewed it as the occupation that would result in the most positive benefit to society.[56] However, Holub's playful, surreal, and subtly ironic commentary extends beyond his poetry and into these same interviews and essays, and as such it is very difficult to find a consistent self-categorisation. While these tactics were no doubt purposely employed by a man who liked to keep his private life private,[57] there are numerous instances when he genuinely appears to contradict himself, and in so doing reveals an identity that is less certain than the previous quotation would have us believe.

For example, in an interview conducted in 1998 with the American author Irene Blair Honeycutt, Holub refers to one of his own essays, and the fact that he is neither a scientist nor a poet:

> As I said in an essay, for 95% of my lifetime I was and am a passer-by, a driver, a lover, a father, a (bad) serviceman, an offender, a shopper, a shadow, and a player. Realizing that, one is compelled to write poetry to save that last 5%.[58]

From this quote, it would appear that Holub valued poetry as his saviour – the one part of his life that was worth saving – and that it is poetry, and not science, that he considers to be of more value, if not on a societal level then at least on a personal one. The Honeycutt interview (1994) and the 'Rampage, or science in poetry' essay (1997) were written within three years of one another at the very end of Holub's life, and so this contradiction cannot be explained by a change of heart caused by a major event such as the Warsaw Pact invasion. However, if Holub did in fact value poetry as a way to save himself, he was under no illusions as to the ethical obligations of poets, noting in the same 1994 interview that:

> Generalizations are misleading. I am not sure that all the glorious statements about poetry refer to the same thing. The 'high ethical imagination' of some poets needs badly a profane counterpart. In our history we have a good example of Jaroslav Vrchlický who wrote highly spirited and most noble poems and, in private, very obscene verses. He had syphilis, anyway. Poetry can prevent evil from happening merely by the fact that when writing or reading a poem one cannot mug somebody – for technical reasons![59]

In the same interview Holub also reveals that while he does not see a difference between a scientist and a poet, he doesn't think that science can inform poetry:

> I don't think science can inform poetry. Practicing science just presupposes a hard-centered approach to any object, including the language. Science presents a firm ground for all personal feelings, a sort of safe existential ground. A scientific worker writing poetry does not see (perceive, feel, sense) the abyss. He is not happy, but he is less desolate and, in many instances, also less neurotic. There is no real distinction in being a poet and a scientist. It is all really part of being human.

Once again, however, he contradicts his own opinion in 'Rampage, or science in poetry', where in the final paragraph he offers some advice on what a scientist is, and the influence that science can have on poetry:

> In conclusion, I suggest (and this is a very scientific form of words) that a scientist, even writing poetry, is and should be and must remain a scientist: science has a deep influence on his personality, giving him assurance and relieving him from abysmal feeling. He is the member of the intellectual community who still has a deep and visible importance for the society and its everyday life. Science is nowadays the hidden and positive 'rampage' of mankind. Science in poetry is not the function of scientific terms, expressions, results,

and technical ideas. It is the function of the scientific, hard-centred approach to reality and to comforting myths. It has its immanent optimism. The poetry of a practising scientist is basically a dialogue; consequently it should be clear enough to be understood and strong enough to lead somewhere in human terms. Science in poetry should shed some relatively new light. It is definitely not the post-romantic and postmodern poetic way of wearing dark glasses on a moonless night.[60]

In his writings Holub often referred to a hard-centred (scientific) and a soft-centred (poetic) approach to making sense of reality, and while he acknowledges the need for both, he is often at pains to point out that he readily adopts the hard-centred approach, and that he is worried that his own work might not be categorised as such. Speaking in the *New Coin Poetry* interview about how science had influenced his poetic writing, Holub notes that:

> I am worried about what can be labelled a 'soft centred' approach.
> I think every fact, every norm, needs some scrutiny, analysis and critique. This I have from science. Because reality, the human, the natural reality, is so complicated I think that in poetry clarity of expression is essential. I don't like poetry which is a jungle of words, of feelings. I need clarity. Seamus Heaney called my poetry the 'fully exposed poem': my poems are fully exposed.[61]

To quote Heaney in full, who was talking about *Sagittal Section*, the first book of Holub's poetry to be published in the United States by Oberlin College in 1980:

> It should be said right away that the book delivers what the title promises – a laying bare of things, not so much the skull beneath the skin, more the brain beneath the skull; the shape of relationships, politics, history; the rhythms of affection and disaffection; the ebb and flow of faith, hope, violence, art.[62]

Holub's choice of Heaney's complimentary words further underlines the point that he was concerned with how his poetry might impact the way in which his science was received. Holub mentions this particular quotation a lot in both his essays and the interviews that he conducted, probably because it is very 'scientific'.[63] While it points to the concrete imagery of his work and implies associations with the influence of Pound and Williams, it also suggests surgical practice, exposing the underlying meaning just as a surgeon would expose the tissue of a patient before beginning to operate. I doubt that Holub would have used a quotation so widely, even one from Heaney, if it had referred to a 'wildly imaginative poem'.

Part of this struggle with his identity was influenced by how Holub felt poets were perceived, not least by himself and his scientific contemporaries. In his conversation with Wilde-Menozzi he discusses the roles that poets play:

> Poets often play roles. Czech poets in these days might end a poem with a question like 'where is my home?' This is not really sincere. They don't want to look hard for answers. Science pushes that, answers leading to other connections – discovering weaknesses, energies, riddles, orders, destructions, work to be done. I see poetry like that too, except that poetry comes into existence and touches existence. Its energy is inexplicable. In a poem like 'Vanishing Lung Syndrome,' I was astonished and taken by the virulent dark energy of the disease. But the poem moves beyond the initial destructiveness. I can't tell you what it means.[64]

In this statement Holub appears to be lambasting his fellow poets for being lazy, and yet at the same time he states that poetry can lead to answers, even if these answers are completely subjective. In the same conversation, he states that he does in fact inform his poetry with science, and that in doing so he exposes the subjectivity that poetry can afford him, noting that: 'I like to colonize poems with words from science. But they work differently. A dog in a scientific paper has one meaning and one meaning only. That dog in a poem can be anything.'[65]

The poem 'Vanishing Lung Syndrome' is taken from his collection with the same name, *Syndrom mizející plíce*, first published in 1990. Vanishing lung syndrome, otherwise known as idiopathic giant bullous emphysema, is a medical condition that typically occurs in young, thin, male smokers. First discovered by Richard Burke in 1937, it is a rare condition that is characterised by damaged alveoli that distend to form exceptionally large air spaces, especially within the uppermost portions of the lungs, compressing and displacing their functional tissue and thereby causing it to appear as if sections of the lungs have disappeared when the region is examined using X-rays.[66] Holub's poem is a surreal description of what might be taking place in the vanishing lung of a patient, in which he colonises his poetry with scientific jargon to present multiple examples of what might be displacing the lung, leaving the reader with multiple interpretations. Holub himself noted that 'The poem is about a rare disease which looks like losing one lung. The lung would be completely replaced by a cavity, and it's some sort of comment on the human soul.'[67]

Although he 'can't tell you what it means', the poem is an example of how Holub used poetry to explore ideas that would otherwise be unavailable to him through science. His poems afforded him the opportunity to 'save' the remaining 95 per cent of his persona, where 'scientist' undoubtedly sat alongside 'driver', 'lover', and 'father'. The final lines of the poem, translated into English by David Young and Dana Hábová, read as follows:

> lost in the landscape
> where only surgeons
> write poems.[68]

These lines might be read as a commentary on what Holub thinks is missing from his soul, that is, if only surgeons were 'allowed' to write poetry he would himself be whole, and no longer 'lost in the landscape'.

While Holub appears to have been in two minds as to whether science and poetry can inform one another, he conceded that they at least have things in common, stating in his conversation with Wilde-Menozzi that:

> Poetry and science have many things in common – discovery and perhaps working on something small, understanding energies,

staying close to facts, repeating and verifying. Even though I am a scientist I can still write a poem. I know what a poem must be to be a poem.[69]

However, in his *New Coin Poetry* interview (conducted in 1996, two years after his conversation with Wilde-Menozzi), Holub states that while there is a common root between the two disciplines, they are otherwise unrelated:

The common root of both is something which is nobly called 'creativity', which I would call the art of getting proper ideas or, more exactly, the proper questions which are very specific for this artistic discipline or for that scientific discipline. Otherwise they have very little in common. Maybe there are some human capacities, faculties which can be used in both; like imagination, like the ability to think analytically.[70]

This is a particularly puzzling statement, especially as Holub also acknowledged on several other occasions that there were times when science and poetry were effectively one and the same. For example, in conversation with the Irish philosopher Richard Kearney, he remarks that 'in the Aristotelian tradition, or in the Greco-Roman tradition, to put it more broadly, it was just one thing – the fable of fables, science alongside types of rhetoric or literature'.[71]

In the same interview, Holub states that the division between the two occurred (in the West) during the first Industrial Revolution, but that it was dependent on the structure and economy of the society in question:

Actually, I would say in a slightly cynical way that the split between science and the arts was pronounced in societies which could *afford* it. With our Czech enlightenment, with its national insurgence at the end of the eighteenth and the beginning of the nineteenth century, there was no such feeling because the nation was fighting for some type of survival. I wouldn't say we were almost exterminated, but the language was retreating into the villages, into the country, and educated language almost didn't exist any more. And under this condition, the redefinition of the nation came in the same way in art, in poetry, and in science. So that science was something which was deeply identified with the national life. But the more we became a modern society, the more we began setting these things apart, which is obviously the 'two cultures' problem; it's a problem of affluent societies, those that can afford it.[72]

As discussed at the beginning of this chapter, this idea of science and poetry developing and proceeding because of, rather than in spite of, adversity is a theme that Holub returns to throughout his various essays and interviews. Indeed, he believed that triumphing over adversity was what defined human history, not just in terms of politics and censorship but also with regard to our own bodies. Writing in *Shedding Life*, the first collection of his essays to be translated into English, published in 1997, Holub remarks:

Although it doesn't sound very uplifting, the basis of the contemporary human being is derived not only from cultural traditions but also from the history of human diseases, which themselves left a pretty strong imprint on cultural traditions and, in addition, influenced our physiology, our immune capacity, and our identity.

The history of life is the history of its being endangered, the endangering of life by life, as much as it is the development and strengthening of life through danger. We can describe human history equally well as a long development or as a long disease.[73]

The personal adversity that Holub faced further reveals the identity with which he wished to be associated, at least in public. While Holub's books were banned and his privileges revoked during the various regimes of censorship that were enforced in Czechoslovakia, he claims that he was never questioned as a poet, only as a scientist:

I was interrogated by the police as a scientific worker, never as a writer. It was scary, since it followed my disclosure of a secret biological warfare project to a British colleague (not that I was audacious; I was simply friendly and found it impossible to lie).[74] Also, in the laboratory, I was once asked to join the Party. (I was the only non-Party member – feeling like a mentally retarded child among all the bright Party members.) But I already had the mental retardation as a life role and simply said that I don't agree with the cultural policy of the Party. This should have been obvious after some surrealistic incidents which I had with the cultural bosses over the years because of my writing. Only recently did I hear that 'my volume' (i.e., report on me, etc.) was the biggest in the Party secretariat in Prague 4.[75]

This statement is reflective of the importance that Holub ascribed to science; that somehow it was only his work as a scientist that was deemed threatening enough to be questioned by the oppressive regime – a justification of the 'work to be done' that was achieved through his science. However, again there are contradictions. Neil Astley, the founder and editor of Bloodaxe Books, quotes Holub in the 'Afterword' to *Before and After* as having said: 'I have a single goal but two ways to reach it. I never switch them – I apply them both in turn. Poetry and science form the basis of my experience.'[76] Similarly, when asked in the Honeycutt interview what he did for fun, Holub replied: 'I do everything for fun! After writing a scientific paper, a poem is fun; and after an aborted poem, looking into the microscope and seeing at least something is great fun.'[77]

Despite his protestations, Holub would appear to have relied equally on both science and poetry. What this statement about fun also reveals is that Holub was perhaps more comfortable in his role as a celebrated scientist than his role as a celebrated poet. Note how he considers poetry to be a reward for scientific success, whereas science is the commiserating arm-around-the shoulder for his poetic failings. This lack of confidence in his poetry is probably reflective of its lack of recognition in the country of his birth during his own lifetime, and appears not to have been countered by the recognition of Holub the poet outside Czechoslovakia; for example, the Poet Laureate, Ted Hughes, once described him as 'one of the half dozen most important poets writing anywhere'.[78]

One reason why Holub made such personal proclamations about being a scientist who wrote poetry in his spare time (and not the other way around) might have been because, despite his defiance as a smiling Sisyphus, he feared censorship from another draconian and regimented regime: the scientific community. As a fierce critic of mysticism and pseudoscience, it is likely that Holub was fearful that he himself would be labelled as such, that is, as the surrealist poet who did some interesting work on immunology, but who was slightly less than credible given his poetic tendencies. As someone who was keenly aware of the split between the two disciplines, Holub's poetry demonstrates that he was able to combine abstract artistry with concrete scientific precision. Perhaps more so than any of the other poets in this book, Holub was widely lauded as an accomplished poet, and yet this was an identity that he himself was never quite comfortable with; in his own mind, he would appear to have been a scientist first and a poet second. Ironically, while his scientific work was important in the field of immunology, especially in understanding the role of the omentum[79] as a record of our immunological history, it is unquestionably his poetry for which he will continue to be remembered.

To conclude this chapter, and in homage to Holub's own treatment of his name, I would like to offer an analogy between Holub and Leaping Lena – a West German racing pigeon that got lost in Czechoslovakia during a routine 1954 flight. When she returned home two days later, there was a message attached to one of her legs addressed to Radio Free Europe, pleading with them to continue their broadcasts and signed simply 'Unbowed Pilsen'.[80] Following this incident, Leaping Lena was brought to the United States and used to raise money for Radio Free Europe as part of a publicity campaign; a symbol of hope from a pigeon that had broken through the Iron Curtain. As previously discussed, 'Holub' translates into English as 'pigeon', and despite the censorship that was imposed upon him by the state and that which he imposed upon himself in reaction to the scientific community, on reading his poetry today there is a clear message of hope; one that breaks through his own iron curtain and that provides us with a message about what humans are capable of if they only try. Leaping Holub – the unbowed pigeon from Pilsen who was always supposed to fly.

6

The poetic pioneer: Rebecca Elson

Before it falls,
And I forget to ask questions,
And only count things.

From 'We Astronomers' by Rebecca Elson[1]

In search of the prehistoric

Rebecca Anne Wood Elson was born in Montreal on 2 January 1960. Unlike the other scientists who feature in this book, there is not yet a large amount of biographical information or essays about Elson, her work, and her life. However, as will become clear, her mastery of both science and poetry means that she is a necessary and welcome addition. Apart from Elson's peer-reviewed articles and published poetry, the majority of the personal detail in this chapter was provided by her family and friends, and taken from the autobiographical essay 'From Stones to Stars', written by Elson in 1998.[2]

Elson grew up in Montreal where her father, John, was Professor of Geology at McGill University. As a young child, her summer holidays were spent roaming around the shores of Lake Agassiz (a prehistoric lake on the border between Canada and the United States), while her father conducted fieldwork, collecting samples of rocks to trace the evolution of the lake's shores, and measuring the roundness of pebbles in search of clues as to what the lake might have looked like in ages past. During these field trips, Elson would help her father in his investigations, poking pebbles through square grids cut into plastic boxes in order to standardise their size and judge their roundness. As well as helping to categorise the stones, Elson also posed in many of her father's scientific photographs of the area, serving as a living scale against which the moraines, banks of clay, and even glaciers could be contextualised and measured. It was during these field trip vacations (which often overran the official school holidays by several weeks) that Elson later acknowledged her scientific education as having truly begun, learning through reflex and osmosis about the methods and principles of scientific research and understanding.

This osmosis continued beyond the trips to Lake Agassiz, as Elson spent her childhood surrounded by scientists and science, from weekend experiments with her father to dinner table conversations with his university colleagues.[3] To Elson and her family, science was not simply something that you did at school, it was a way of life, and her inquisitive nature was celebrated and encouraged by her family and their circle of friends. Elson continued her extra-curricular exploration of science by taking part in treasure hunts at McGill's Redpath Museum, and by visiting the annual McGill open houses, where children and adults alike could roam the halls of various science departments, observing giant working models of volcanoes and other kinaesthetic tools that brought science to life.

Despite always being top of her class at both elementary and high school, Elson had a rather negative opinion of the way that science was taught during her formal education. Like Davy and Ross, Elson considered school science to be rote learning from textbooks that stifled rather than encouraged inquiry and investigation, remarking that: 'The subject may have been science, but the process wasn't.'[4] Her father and mother played an important role in Elson's educational development, demonstrating to her that education didn't just come from school, and that you were not limited to only being interested in a single subject or discipline. Several of her poems refer explicitly to the positive influence that her father in particular had on her scientific endeavours; her admiration and respect for her father's encouragement are captured in the final lines of 'Poem for my Father', in which she makes clear that these early scientific adventures were the first steps towards her later scientific accomplishments:

Like the first hominids,
Our footprints trailing out behind,
You honouring all my questions
With your own.[5]

In high school Elson was enrolled in a bilingual English and French programme, and in addition to excelling in her academic subjects she liked to sew and cook, starting a small catering business with a friend that made hors d'oeuvres for parties and using the skills that she was taught by her mother to make evening dresses for parties and other social occasions. Her dedication to a variety of interests was encouraged by her parents, who themselves hosted amateur music and play-reading evenings for their friends, encouraging Elson and her sister Sarah to participate as well as watch. Her childhood also included many visits to the ballet and musical concerts, and from an early age Elson's parents read broadly to both children; her mother, Jeanne, had previously worked as a librarian at Yale University, and would often regale her children with whatever it was that she happened to be reading at the time. These readings included poetry, Elson's earliest experience of which was when her father read 'A Child's Christmas in Wales' by Dylan Thomas to them during the festive season. As well as listening to and reading poetry, Elson also began to write poems from a young age, her first forays surfacing alongside those initial scientific endeavours: in the camper van while driving along the shores of Lake Agassiz. Elson's English teachers further encouraged her writing, and some of her first poems were published in the elementary school newsletter.

Elson's earliest scientific heroes included the English anthropologist Jane Goodall and the French explorer and conservationist Jacques Cousteau (she was especially interested in the exotic fauna of the Galapagos Islands that he brought to life through his writings), and it was this broad exposure to a range of scientists and sciences that led her later to remark that: 'It was never the facts that interested me so much as the possibilities they opened up to the imagination.'[6] Elson's childhood also coincided with a time of extraordinary space exploration, including the moment when Neil Armstrong stepped on to the surface of the moon on 20 July 1969. These events further inspired Elson, and her fascination with astronomy stemmed from when she would stare out of her window at the skies of northern Canada, at the stars up above and the spaces in between.

After finishing high school, Elson was encouraged by her parents to attend university in the United States, and following the recommendation of a close family friend she enrolled at Smith College in 1976 at the age of 16.[7] Initially she had considered an English major, but she came to the realisation that she could pursue reading and writing in her spare time; her love of the Galapagos Islands combined with a late high school fascination with genetics encouraging her to instead pursue a biology major.

From ice cream to isolation

Despite her initial decision to major in biology, Elson never took a single class in the subject, instead switching to astronomy after receiving a recommendation from a friend. One of her astronomy teachers was the German astronomer Waltraut Seitter, who would later go on to be the first woman in Germany to hold a chair in astronomy. At times some of the astronomy class would pile into Seitter's old VW Beetle and go downtown for ice cream, trips that were accompanied by discussions about the mysteries of the universe, where once again Elson was encouraged to ask questions. During her time at Smith, Elson also spent many nights at the McConnell Rooftop Observatory at the Clark Science Center, learning how to use the small telescope to find objects in the night sky. Occasionally the class would also head to another observatory at one of Smith College's field stations, located in the nearby town of Whatley. Here, Elson could use the larger, 16-inch telescope to further explore the universe, and it was at this site that she saw for the first time the Milky Way's sister galaxy, Andromeda, which left a lasting impression on her.

While studying at Smith, Elson spent her junior year abroad at the University of St Andrews in Scotland, reflecting on her time there in the first draft of a semi-autobiographical poem:

> To a tiny Scottish town, cathedral ruins
> Castle ruins, wet sands, east sands, fishing harbour
> Cold sea licking & sucking, rain, all-day sunsets.[8]

While in Scotland, she was also encouraged by one of her Smith tutors to apply for a summer placement as a research assistant at the Royal Observatory of Edinburgh, working with the South African astronomer Tim Hawarden. Elson later described this as one of her most inspirational experiences in becoming an astronomer, and during that summer she spent her time carefully observing 935 glass plates that had been produced using the Schmidt telescope at Caltech's Palomar Observatory in the 1950s and 1960s. The plates, which were approximately 14 inches long on each side, represented a map of the entire northern celestial hemisphere, and Elson was tasked with looking through them in search of elliptical galaxies that were bisected by a line of dust. Elliptical galaxies were thought to be representative of ancient systems, meaning that almost all of the dust and gas would have been used up in the creation and formation of stars; as such, elliptical discs that contained strips of dust represented something of an irregularity.

Once Elson had finished mapping out the northern hemisphere (a task that was done using only a hand lens), she moved on to the southern celestial hemisphere, using plates that had recently been produced by the UK Schmidt telescope at the Siding Spring Observatory near Coonabarabran, New South Wales, Australia.[9] This research was later published in the *Monthly Notices of the Royal Astronomical Society*, with Elson listed as second author.[10] Despite what many would consider to be a monotonous task (scanning glass plates of the entire sky with a hand lens in search of rare galaxy formations), Elson found herself enamoured by the research, losing herself in the plates' depiction of the universe, just as she had when looking out at the Canadian night sky as a child.

After her experiences in Scotland, Elson was determined to pursue a career in astronomy, and was also set on returning to the UK to continue her studies. However, she knew that in order to do a PhD in astronomy she would need to demonstrate a greater understanding of physics than had been afforded to her at Smith. After graduating in 1980, she therefore returned to Canada, winning a scholarship to undertake a Master's degree in physics at the University of British Columbia (UBC). Elson initially struggled at UBC, finding large sections of the course to be at times overwhelming, which was unsurprising given the relative limits of her education in the subject at Smith. Furthermore, in a similar way to Maxwell and his frustrations with the Tripos, Elson found many of the exercises completely irrelevant to actual research and scientific inquiry; why, for example, would she ever need to calculate the trajectory of a bubble in an imaginary room of glass? Following poor grades her scholarship was withdrawn in the second year of her studies, and so she supported herself by working part-time as a laboratory assistant. Despite the difficulty, and at times irrelevance, of the degree programme, Elson found solace in the fields of quantum mechanics and general relativity, the former presenting intriguing paradoxes for her to ponder over while the latter allowed her to further probe the limits of the universe.

In addition to the demands placed on her by the course material, Elson also found herself in a patriarchal environment, which at times she found to be alienating. Her undergraduate programme had been made up entirely of women, but at UBC she often found herself the only female in a class or seminar, describing the experience as often feeling as if she had walked 'into the men's bathroom by mistake'.[11] Despite these barriers, Elson's grades improved and she successfully applied for a PhD in astronomy at the University of Cambridge. After handing in her Master's dissertation, she flew to the UK in 1982 to begin her doctoral research, supported by a prestigious Isaac Newton Studentship.

At Cambridge she studied under the English astronomer Michael Fall, investigating clusters of stars in the Large Magellanic Cloud (LMC), the third-closest galaxy to the Milky Way. The LMC contains several clusters of young stars, and at that time it was uncertain why this should be the case, given that similar clusters that had been observed in the Milky Way had been found to be far more ancient. During her PhD, Elson also spent time at the Mount Stromlo Observatory in Canberra working with the Australian astronomer Ken Freeman; it was here that she first observed for herself the LMC, which is only visible as a faint cloud in the southern celestial hemisphere and from latitudes south of 20°N. Some of the expectations of these observations are captured in her poem 'Observing', the first stanza of which recalls the excitement and almost hushed reverence of what it meant to be watching the stars while everybody else slept:

At the zenith of the night,
Becalmed near sleep
In your dark blind of dome,
You hear it move.[12]

After returning from Australia, Elson learnt that Fall had taken up a position at the headquarters of the Hubble Space Telescope Institute, in Baltimore, Maryland. Not wanting to lose the continuity that Fall had provided and sensing an opportunity to integrate herself within the blossoming Hubble Space Telescope community, Elson kept Fall on as her primary supervisor, remaining in Cambridge but visiting him regularly in Baltimore.[13] While Elson enjoyed the intellectual stimulation provided by these frequent trips aboard, she also found them to be lonely and at times isolating.

In contrast to the isolation that she found abroad, she greatly enjoyed the time that she spent in Cambridge, returning between trips to the house that she shared with her friend, the historian Petà Dunstan, who was at the time studying for her PhD in the Faculty of Divinity. During this time Elson also continued to enjoy reading and writing, although at the start of her PhD she was mainly focused on creating narrative fiction, short stories, and song lyrics, influenced by the lyrics of Leonard Cohen and Jacques Brel, but also those of traditional Irish and American folk songs. As the frequency and length of her trips abroad increased, however, she began to turn her creative output back towards the poetry of her youth, often sending poems to Dunstan during the regular correspondence that they enjoyed, which helped to combat Elson's isolation during these extended trips. She became a voracious reader of poetry, enjoying the works of Robert Frost, Dylan Thomas, Anna Akhmatova, and Maya Angelou, but also Virgil, Dante, Shakespeare, and Coleridge, and like her librarian mother she could often be found with a book either in her hand or on her knee. Elson published several papers during her PhD, and her work on the age calibration and age distribution for rich star clusters in the LMC remains a milestone in the field.[14] By mapping the distribution of cluster binaries, studying stellar mass segregation, and obtaining the distribution of stellar luminosities to very faint magnitudes, she established fundamental connections between the evolution of individual stars and the dynamics of the clusters as a whole, and in doing so helped to develop a better understanding of the classification of stars and their systems.[15]

Elson completed her PhD in 1986, after which she was awarded a research fellowship at the Institute for Advanced Study (IAS), an independent, postdoctoral research centre for theoretical research and intellectual inquiry in Princeton, New Jersey. However, unlike the period that she had spent at Cambridge, her time at the IAS was beset by frustration. She had originally been taken on to work with the new Hubble Space Telescope data that was expected to be produced later that year, but the *Challenger* disaster of 1986 brought the NASA space programme to a halt, postponing the launch of the telescope for several years.[16] As one of the few observationalists at the IAS, Elson now found herself surrounded by theoreticians, who considered their work to be superior to hers, and in an environment in which 'Complicated equations would be solved on blackboards, chalk dust flying, in much the way other species might beat their chests or erect gaudy plumage.'[17]

As had occurred at UBC, Elson found her situation exacerbated by the irretrievable 'maleness' of the IAS, with women not only in the overwhelming minority, but also bullied and chastised by their male counterparts.[18] Sexist remarks and jokes were the norm, with any female colleague who failed to laugh being accused of lacking a sense of humour. This hostility towards women was discussed by the Northern Irish astrophysicist Jocelyn Bell Burnell in her presidential address to the Royal Astronomical Society in 2004, in which she made specific reference to the difficulties that Elson had faced while in Princeton:

> I spent time in Princeton University's Physics Department and the women's bathrooms in that department were remarkable. They were the men's bathrooms with only the name on the outside of the door changed! What a brilliant way to make women feel honoured and welcome members of your department.[19]

Despite the personal and professional hardships that Elson endured while at the IAS, she continued to pursue her research by obtaining extensive observations of the globular cluster system in the LMC, using ground-based telescopes while the wait for data from Hubble continued. In 1987, during her first year at the Institute for Advanced Study, she wrote a definitive article on the dynamical evolution of globular clusters for the *Annual Review of Astronomy and Astrophysics*, becoming one of the youngest authors in that publication's history, and creating a work that is still regularly cited.[20] After three years at Princeton, however, Elson felt that her career was at its lowest ebb and considered herself to be an outsider among her scientific contemporaries. Her only relief came from the small poetry groups that she hosted on Tuesday evenings in her apartment, where like-minded souls would read and share poetry in a supportive and constructive environment; a stark contrast to the research community from which she continued to find herself isolated.

With the support of her poetic contemporaries, Elson began to publish her poetry while at Princeton, and it would later appear in periodicals and magazines in both the UK and the United States, including *Poetry* and *The Rialto*. Elson developed many of her poems from notes that she kept in thick, hardback, A4 notebooks. She would write in pencil, and many of these notes and drafts, some of which she would later re-edit and form into poems, present a raw and insightful vision of how Elson viewed the world around her. Some of the entries from these notebooks are reproduced in *A Responsibility to Awe* and *Oxford Poets 2000: An Anthology*.[21]

In 1989 Elson left Princeton to become a fellow at the Bunting Institute at Radcliffe College.[22] As well as continuing her work on globular clusters, in 1989 Elson also became the youngest astronomer to serve on a committee for the United States National Academy of Sciences decennial review. However, Elson once again found herself frustrated with the systems and cliques of university research environments, for while the Bunting Institute provided her with a stipend and access to resources, she still needed independent affiliations with a local laboratory, something that proved to be very difficult with the Harvard Center for Astrophysics.[23] Elson was only at the Bunting Institute for a year, but by the end of her fellowship her experiences, combined with those at the IAS, had left her so disillusioned with the astronomical research community that she very seriously considered giving up on her career as an astronomer. Returning to her love of writing she applied for a teaching job with the Harvard Expository Writing Program, which would involve teaching essay writing to Harvard undergraduate students. She was ready to join the creative writing community there when she received an offer from the Institute of Astronomy in Cambridge to work on data from the soon-to-be-launched Hubble Space Telescope. This was a renewable postdoctoral position that she didn't feel she could refuse, and so after teaching for a term on the Expository Writing Program she returned to Cambridge in the spring of 1991.

Aberrations

After several years of delays, brought about by issues with funding and the *Challenger* tragedy, the Hubble Space Telescope was successfully launched into an Earth orbit on 24 April 1990 on the Space Shuttle *Discovery* during the STS-31 mission. The original budget for the telescope has been $400 million, but this had increased to $2.5 billion by the time of launch, making it the most expensive scientific instrument ever assembled at the time. These exorbitant costs were in part because of the need to keep the telescope in a cleanroom for the four years between the planned and actual launch date, at a cost of approximately $6 million per month.[24] Given these costs, there was a lot of pressure on NASA and the European Space Agency (which had jointly funded the mission), as well as all of the associated researchers, to deliver spectacular results. With a 2.4-metre mirror, the four instruments on board the Hubble Space Telescope were designed to take measurements in the near infrared, visible, and near ultraviolet parts of the electromagnetic spectrum, staring through time and space to provide a better understanding of the formation of the universe. It was hoped that the initial images produced by the Hubble Space Telescope would not only capture the imagination of the world's media, but also revitalise the deteriorating relationship between the space programme and the general public, especially in the United States.

Within weeks of being launched the Hubble Space Telescope began to return images of distant star systems. However, while these images appeared to be sharper than those that had previously been achieved using ground-based telescopes, they were dramatically lower in quality than had originally been expected. An analysis of these flawed images revealed that the main mirror had been polished to the incorrect shape, and that at the perimeter it was too flat by approximately 2.2 micrometres, approximately 1/50th of the width of a human hair.[25] This aberration meant that while the telescope could still create high-resolution images of bright objects, the faint and distant objects that Hubble had been primarily designed to observe could not be measured to the degree

161

of accuracy that was required by astronomers who were working on these objects, among them Elson and her continuing research on globular clusters. In her poem 'Aberration', Elson explores the frustrations of coming so close to the desired goal:

> And just for a moment
> You forget
> The error and the crimped
> Paths of light
> And you see it[26]

In this poem, Elson concisely expresses the frustrations that must have been felt by both her and her fellow astronomers; the detailed answers to their questions frustratingly out of reach by less than a 1/50th of the width of a single strand of hair. Despite these frustrations, however, Elson remained optimistic, the rest of the poem reading as a love letter to the potential of what can be achieved through science.

This optimism is typical of most of Elson's poetry, both in her finished poems and in the poetic notes and drafts that she kept in her journals. In reading and re-reading Elson's poems, this optimism manifests itself as a kind of scientific pragmatism; a certainty that comes from someone who knows that the answers are out there, and that you just need to be willing to find them. In addition to this optimistic pragmatism, her writing is characterised by the clarity with which she is able to explain incredibly complex topics, creating phrases and images that are not just scientifically accurate but also poetically poignant and relatable. The following lines are taken from a note dated 19 February 1998, shortly after Elson had attended a lecture on string theory:

> We are only seeing things from different points of view
> In certain limits a theory metamorphoses
> Into another, and another
> Space curving in on itself[27]

162

String theory is sometimes called the Theory of Everything, as its ultimate goal is to unify quantum mechanics with general relativity, and in so doing explain all of the particles and forces observed in nature. String theory predicts a universe made up of more than the four dimensions (x, y, z, and time) that we are used to observing in our everyday lives, with proponents of the theory arguing that these extra dimensions are folded in on themselves, or 'compactified', rendering them unobservable in a four-dimensional world. These lines from Elson's poem create a vibrant and distinct image of this extremely complex topic, not only clarifying and helping us to visualise this compactification but explaining that even if we have not yet been able to observe these extra dimensions, it is not because they are not there. For a scientist who primarily classified herself as an observationalist rather than a theorist, this leap of faith is emblematic of the optimistic pragmatism that permeates her work and writing.

Another example of the clarity and pragmatism that permeates Elson's poetry can be found in the poem 'Let There Always be Light'. Here, Elson writes about the search for dark matter, the non-luminous material that is postulated to exist in space and yet is not directly observable. This 'missing' matter is thought to play a key role in the fate of the universe, which, while currently expanding, will either continue to expand (an open universe scenario colloquially known as the 'Big Freeze') or else will contract (a closed universe, in which the current expansion will eventually stop, before collapsing back in on itself into a 'Big Crunch'). In the poem, Elson talks about her own observations of dark matter, and her hopes for a closed universe scenario:

> Let there be enough to bring it back
> From its own edges,
> To bring us all so close that we ignite
> The bright spark of resurrection.[28]

Elson once again writes with a characteristic pragmatism that is underlined by an optimistic faith in a favourable result. The pragmatism of her scientific endeavours suggests to us that this dark matter[29] (whatever it may prove to be) will eventually be found, while her optimism makes it clear that she would prefer a scenario in which there is enough dark matter to ensure that in the end we are united, preferring the 'bright spark of resurrection' to the vast emptiness of an open and ever-expanding universe.

This optimistic pragmatism further reveals itself in her autobiographical essay 'From Stones to Stars', in which Elson writes about the aberrations of the Hubble Space Telescope, focusing on the opportunities, rather than the challenges, that it presented:

> Astronomers using the telescope persevered, applying fancy mathematical algorithms to sharpen up the images (the same technique as might be applied to sharpen up a photo of the license plate of a distant speeding car). Even with the focus problem, the detail in the images was stunning compared to what was possible from the ground.[30]

The error caused by the aberration was both stable and well characterised, meaning that, as Elson describes, it could be corrected for. During this time of aberration, Elson published several significant pieces of scientific research, including work on star clusters in the LMC, and an interpretation of some of the first results to come out of the Hubble Space Telescope.[31] However, despite the research that was being produced by Elson and others, to the majority of the general public and the media the Hubble Space Telescope was seen as an expensive failure, and something needed to be done to rectify this situation. Thankfully, due to the limited lifetime of certain parts, the Hubble Space Telescope had been designed to be serviced by astronauts, and so a fix could be both planned and implemented.

In 1993 a servicing mission was launched to correct the mirror to its intended quality. During this mission the Corrective Optics Space Telescope Axial Replacement (COSTAR)[32] was used to correct the spherical aberration; the Wide Field and Planetary Camera 2 (WFPC2) was also installed at this time, replacing the telescope's original Wide Field and Planetary Camera. The completed repairs and upgrades were a phenomenal success, with the COSTAR correcting for the aberration in the Hubble Space Telescope's other instruments, and the WFPC2 producing spectacular pictures that rekindled the global public's interest in the mission, turning it from a failure into a celebrated success.[33] As the space historian Robert Zimmerman put it: 'It was as if the human race had been living in a fog, and that fog was suddenly lifted, revealing the heavens in all their glory.'[34]

After the telescope was fixed, Elson was part of several international collaborations that used these sharper images to better understand the mysteries of outer space, and in the early to mid-1990s she authored and co-authored several influential scientific studies, including the determination of structural parameters for globular clusters outside the Local Group (the galaxy group that includes the Milky Way), and the interpretation of results from the Medium Deep Survey, one of the three key projects in the early lifetime of the Hubble Space Telescope, which was used to collect hundreds of images in order to explore changes in galaxy shapes over time.[35] In addition to her work with the Medium Deep Survey and globular clusters, Elson also used stellar counts to help set strong limits on the amount of dark matter within the Milky Way, and in doing so she began to answer the questions that she had posed through her poetry.[36]

While publishing regularly in scientific journals, Elson also continued to create poetry, joining a regular writing group in Cambridge where she was encouraged to publish her work in magazines and journals, gaining particular success with *The Rialto* and being invited to read at Simon Armitage's masterclass at the 1998 Aldeburgh Poetry Festival. However, despite being a published author, Elson was reluctant to share her writing with those outside her poetry group, and while she would occasionally show it to her family and friends, she refrained from sharing it with any of her scientific colleagues, perhaps hesitant of the reception that she might receive from some quarters as a young female academic who was trying to make a name for herself in an overtly patriarchal field. On the few occasions when she had shared her poetry with her scientific mentors, they were at best dismissive, with many of them considering her writing to be a side-show to her scientific work. Perhaps unable to relate to an approach that was different to the traditionally academic, single-minded intellectual pursuit of a particular field, they saw her creative writing as a distraction from her professional work, with one supervisor even calling into question her 'commitment' to a career in science as a consequence. While this meant that Elson stopped sharing her poetry with such colleagues, she refused to stop writing, and instead considered the creative freedoms that her poetry afforded her to be paramount to the development of her intellectual curiosity and her capability in the field of astronomy.

While many of Elson's poems deal with science and her love of the subject, she did not write exclusively about science, as her friend Petà Dunstan notes:

> The tendency of some scientists to pare everything down to constituent parts and accept nothing that could not be 'proved' was alien to her. I would contend that the power of her writing is that it can connect at many levels. It is not 'scientific' poetry. It is poetry that uses scientific themes – but also there is much more there too. She used science as a theme not because she wanted to be a poet of science but because science was a large part of her daily life – but so were many other things. And all came into her writing. That is why she can appeal to a variety of readers on a variety of levels.[37]

Those other things included cooking, football, craftwork, gardening, DIY, playing the mandolin, speaking three languages (English, French, and Italian), and spending time with her husband, the Italian artist Angelo di Cintio, whom she married in 1996, having met him at Cambridge through mutual friends in 1983 while she was studying for her PhD. They kept in touch through the years despite often living in different countries, and eventually di Cintio moved to Cambridge in 1993 so that the two of them could be together. They enjoyed spending their time travelling to

Italy, France, and the United States, and several of Elson's poems are about these adventures and their time together; the last verse of the poem 'Hanging Out his Boxer Shorts to Dry' playfully capturing the optimistic pragmatism that so characterises her work, the laborious yet rewarding task of observation replaced by the mundane yet fulfilling sanctity of marriage:

> I love to set them sailing out
> All down the garden,
> My private regatta,
> My flags of surrender.[38]

That Elson was able to fit so much into this short period of her life is made even more remarkable by the fact that in 1990, at the age of just 29, she was diagnosed with non-Hodgkin lymphoma.[39] With treatment the cancer went into remission, yet despite initial grounds for optimism it came back. In her notebooks and through her poetry, Elson talked about the cancer and its effect on her life with openness and sincerity. In her notebooks she describes the cancer with the same vivid imagery that characterises almost all of her poetry; for example, in an entry dated 17 August 1995 she wrote:

> Those small cells already hiding out
> In the Aladdin's cave of your bones
>
> And those small cells already
> Lighting their fires
> In the genie caves of my bones
>
> And the blood sowing its seed   its red seed
> In the dark earth of my bones[40]

There is an inevitability to her writing; however, despite this there is rarely anger or resentment in Elson's poetry, but rather an effort to come to terms with her illness and the attempts at treatment. In the poem 'In Me Now', she writes how:

167

In me now
Are traces of the Madagascar periwinkle
Mustard gas
And mutant genes
And things made inside mice[41]

The Madagascar periwinkle is a species of flowering plant that is native to Madagascar, where for centuries traditional healers have used it to treat medical conditions such as diabetes. These treatments led to its study by Western scientists, who then discovered that extracts from the plant were found to be active against lymphocytic leukaemia in mice.[42] Similarly, in the course of developing chemical weapons and protective agents against them during the Second World War, researchers in the US Army discovered that the nitrogen mustard compound could be used to treat lymphoma.[43]

This poem demonstrates how scientifically literate Elson was with regard to both her illness and its treatment, and many of the poems that she wrote about her illness explore her condition through the lens of an observationalist. In a poem recorded in one of her notebooks on 9 October 1998, entitled 'Transumanza', she writes:

Is there any language, logic
Any algebra where death is not
The tragedy it seems
A geometry that makes it look
Alright to die
Where can it be proved
Some theorem
If P then Q and all is well
If not P then not Q either and all is gone
Or if not P then Q[44]

These last three lines are reminiscent of the equations that are used to represent Bayes's theorem, which describes how the conditional probability for an observed outcome can be computed from knowing the probability of each cause and the conditional probability of the outcome of each of these causes. Elson used Bayesian statistics in her work classifying galaxies,[45] and in this poem she questions whether a similar approach might also be used to create a situation in which her death is not classified as a tragedy, but is rather seen as something else entirely, the 'something else' to be revealed by solving the statistical equations. In a much earlier notebook entry, written on 7 August 1995, Elson also uses her astronomical work to provide an analogy for her illness, this time comparing the way in which she is studied and examined by her doctors to the way in which she herself studied the photographic plates of distant galaxies:

> how it feels to be a photographic plate of the Universe, full of
> galaxies
> being scanned, the xray machine, like a telescope, and me bathed
> in
> xrays, like standing space in front of a pulsar spraying xrays into
> its sky[46]

Despite this poetic and scientific exploration of her condition and its treatment, she was not under any illusions about its brutality, writing on 28 February 1999 that:

> There is no poetry to cancer
> To the body betraying itself
> Ravishing itself
> Leaving itself drained[47]

Similarly, she refused to be defined by her illness, writing on 10 May 1993 that: 'The thing is not to let the doctors take the poetry out of your body.'[48] Throughout her illness she refused to have the poetry taken out of her body, continuing with her writing and her research, and all of the other things that she loved. In her final published poem, 'Antidotes to Fear of Death', which appeared in the summer 1998 issue of the *Orbis International Literary Journal*, Elson writes how:

Sometimes as an antidote
To fear of death,
I eat the stars.

[...]

No outer space, just space,
The light of all the not yet stars
Drifting like a bright mist,
And all of us, and everything
Already there
But unconstrained by form.[49]

In these verses, Elson continues to display the optimistic pragmatism that so characterises her writing – yes, she will die, but in doing so she will return to the stars from which she was born, those bright spots of light that she loved to observe, either by looking out at the ethereal skies of Canada or by poring over photographic plates with a hand lens in a small Scottish town.

Rebecca Elson died on 19 May 1999, at the age of 39, and after her death a horse chestnut tree was planted in the grounds of the Institute of Astronomy in Cambridge in her memory. A volume of the poetry that she wrote from her teens until shortly before her death was published posthumously as *A Responsibility to Awe* in 2001, edited by di Cintio and her friend, the poet Anne Berkeley, who had also been a member of the same Cambridge poetry group as Elson.

Breaking the silence

Despite the obstacles and barriers that she faced in her life, Elson never did choose between science and poetry. In a poetic note dated 11 May 1997 she states that:

Science is not what they say, so serious
The truth being what you imagine
Not what you see

And not something useful
Or something that pays[50]

She saw herself as a rounded person who studied and researched astronomy, and who also wrote poetry. The facts and theories of research were one aspect of her persona, just as the vivid imagination of creative writing was another; but rather than separate compartments, they were all part of her one life, a life that was also filled with language, and sport, and crafts, and family, and everything else from which she took pleasure. To Elson, science was poetry and poetry was science; they were not mutually exclusive entities but rather small facets of existence that helped to fill in the bigger picture of the universe and her place within it.

In terms of her scientific achievements, Elson is today best known for her work on the structure and luminosity functions of rich star clusters, especially the clusters of the LMC and the Milky Way galaxy, and right until the very end of her life she continued to extend this work through the use of ground observations and images from the Hubble Space Telescope.[51] She published over fifty professional works during her academic career, and that many of these continue to be cited today indicates the scale of her contribution to the field. The sexism that she faced as a female astronomer in the 1980s and 1990s meant that she was understandably hesitant about sharing her poetry with her scientific contemporaries, but she was proud to be a pioneer for women in the field of astrophysics, and before she became seriously ill she had begun to research a book on women astronomers through the ages.[52]

Through her poetry, Elson is revealed to be an optimistic pragmatist, one for whom life's many challenges were not viewed as a nuisance, but rather as a great opportunity. Her poetry also reveals the extent to which she loved life, and the bravery and strength with which she faced these challenges, including her battle against cancer. Her husband Angelo di Cintio recalls the following incident that occurred towards the end of her life, which is demonstrative of Elson's strength, courage, and also acceptance:

A month before she passed away, she asked me to take her on the Cam on boat, not a punt, towards Grantchester, where the river meanders through meadows and fields, coasted by willows, some old, tall and broken, some small spiky and vigorous. A grey afternoon but not rainy we went; she was on palliatives, painkillers, steroids. Half way there she told me she wanted to row the boat herself. She rowed the blue boat all the way to Grantchester, staring at the river, as it flew away from her blue eyes. The river of ducks and water grasses, of swans and clouds and ripples. We didn't come across anybody else on that stretch of the river, for an hour or more I was the only witness to the silent fury of her rowing, relentless and scream like, raging against everything and all, and I can't say how deep or high the list of targets for that rage went. When we arrived, her eyes went shut. No tears. Instead her arms and legs were shaking gently. They were doing the sobbing. They were doing a pattern of knocks on the floor of the boat. A slowing down repetition. Two ducks landed on the water nearby and the same pattern was there, extended in distance. The expression on Rebecca's face was empty, and luminous. This was the only time I could call a breaking down in Rebecca's last year. And yet perhaps it wasn't. Perhaps it was another curve in the form of her 'allowing'.[53]

Despite an understandable reluctance to share her poetry more widely among her scientific contemporaries, Elson was someone who inspired poetry in others. An online article for the journal *Nature*, which was written about Elson's discovery of a white dwarf in a young star cluster in the LMC,[54] discussed the importance of this research in terms of its scientific credentials, concluding with the following observation: 'It is often said that we are stardust; but we are still learning what kinds of stars made us.'[55] Similarly, in an obituary that appeared in *Physics Today*, the astronomers David Helfand and Gerry Gilmore (a long-time collaborator of Elson) wrote: 'Becky's scientific contributions helped significantly in widening our horizon, and her life made the universe just a bit more glorious.'[56]

In reading Elson's poetry, we are reminded of the necessity for science and poetry to coexist and interact with one another, rather than to be consigned to uncommunicative and isolated boxes. In one of her draft poems, entitled 'How Science Works', Elson begins with a quotation from the Austrian philosopher Ludwig Wittgenstein: 'What we cannot talk about must be consigned to silence.'[57] Elson knew that science and poetry were only two parts of a greater whole, and that both needed to be considered in order to better make sense of the universe and our place in it. Working exclusively within the silos of poetry or science restricts our language, widening the spaces between us and making the silence deafening. Elson knew this, and through her simultaneous pursuit of both scientific and poetic understanding she was able to break through this silence, bringing us closer to that 'bright spark of resurrection'. By reading her poetry and the life that it contains, we are reminded of why we should never stop either asking questions or pursuing their answers.

Epilogue

Science explains nothing
but holds all together as
many things as it can count

From 'Science' by Robert Kelly[1]

In the Introduction, I laid out my aspirations for this book: that it
would provide a tentative explanation as to why the scientists that
I had selected wrote poetry, and how their individual styles and
reasons for writing verse might have concurred. Despite the differences in style and approach of each of the scientists that feature
in this book, I believe that they were in part able to achieve their
scientific feats because of their interest, talent, and enthusiasm for
poetry. From Lovelace's metaphysical insight through to Holub's
precise imagery, poetry was an integral part of their personal and
scientific identities; a tool that enabled them to better understand
how their science could provide answers to the questions that they
explored. Elson perfectly captured this sentiment in her poem 'We
Astronomers':

Before it falls,
And I forget to ask questions,
And only count things.[2]

Elson's thoughts are echoed in the extract from Robert Kelly's
poem 'Science' that begins this chapter, and both highlight the
danger of failing to see the bigger picture by becoming lost in the
miniature. For all of the scientists featured here, their ability to

embrace both science and poetry gave them an aptitude to better consider the scale of what they sought to achieve. Yet despite their numerous successes in both fields, one thing that unites all of the poetic scientists in this book is that at various points in their lives external forces wanted them to choose either science or poetry – from Davy's assertions to his mother that she ought 'not to suppose I am turned poet' through to Elson's decision to keep her poetic talent largely hidden so that she would be taken more 'seriously' as a scientific researcher. Even Holub, whose poetic achievements will probably far outlast his scientific accomplishments, considered himself to be a scientist first and a poet second. But why did they have to choose?

I argued earlier that Holub made this decision because he feared alienation from the scientific community. This fear of alienation no doubt lingered for the other scientists in this book; even Ross, who published a relatively large selection of poetry during his lifetime, constantly felt the struggle between 'Science' and 'Thought'. Despite the symbiotic nature of poetry and science that is evidenced from their accomplishments, these poetic scientists struggled to accept the interdisciplinarity of their approach. However, this struggle is not unique to these scientists alone, and I believe that it is symptomatic of our relative failure (in the Western world at least) to truly embrace the potential for interdisciplinary research.

We will only be able to solve complex problems, such as global warming or societal injustice, by first recognising that these issues are interdisciplinary in their nature, and by realising that in order to solve them we need to work together, across the disciplines, to create new solutions that are far greater than the sum of their parts. However, a refusal to consider opportunities for interdisciplinary thought is ingrained in us from an early age. In the UK, the national curriculum is simply not set up to promote interdisciplinarity, neatly carving education and learning into different subjects, which are taught almost universally as mutually exclusive entities. If we are taught from an early age that each of these subjects has different approaches, solutions, and limitations, without ever being encouraged to use the approaches from one discipline to address the limitations of another, then what hope is there for an environment in which interdisciplinary problems are even recognised, let alone solved?

Medicine has been said to be both a science and an art.[3] However, how many schoolchildren are taught that the critical awareness, reflection, and decision-making skills that are so fundamental to being a successful artist are also proficiencies that are required of consultants and surgeons? Similarly, how many maths teachers inform their students that the ability to manipulate precise and intricate ideas while formulating a methodology to describe the physical world are aptitudes that many successful poets would identify with? By encouraging students (from pre-school to postgraduate) to operate inside these boxes, we are fitting them with blinkers that restrict the scope of their vision and the application of their potential talents. And in so doing, we are effectively forcing people to choose at an early age whether they consider themselves to be a poet or a scientist, rather than demonstrating that these two disciplines offer complementary ways of analysing and describing the world and its surroundings. This is not to say that science and poetry will always agree, but rather that any incongruity merits further investigation.

In his 'Defence of Poetry' the English Romantic poet P. B. Shelley noted that: 'Poetry is a mirror which makes beautiful that which is distorted.'[4] By looking in this mirror, scientists are presented with an alternative worldview, and with it perhaps a different appreciation and understanding of the scale of their achievements. Utilising this mirror demonstrates that it is simply not the case that scientists cannot 'do poetry', or that poets cannot 'do science', as provided with the right training, experience, and practice, science and poetry can be 'done' be anyone. However, given the current constraints that are placed on both professional scientists and poets in terms of their research and practice (as well as their various teaching and administrative duties), it is very difficult to find the time and energy needed to develop and hone a craft in either discipline, let alone both of them simultaneously. In which case, interdisciplinarity can still be achieved, albeit through meaningful collaboration.

In the Introduction I talked about 'Peer Reviewed Poetry', the spoken word show that inspired the research that underpins this book. This show was by its very nature dualistic, and arguably contributed to a separation of the two disciplines rather than exploring the potential for interdisciplinary collaboration. Dan and I have since developed a new spoken word experience, one that aims to explore and celebrate the collaborative nature of scientists and poets. 'Experimental Words' is a project that has been funded by the UK's Arts Council, through which a selection of poets and scientists have been paired up; via facilitated workshops and mentoring these pairings are then encouraged to work together to create original spoken word pieces which are then performed to the general public. Through 'Experimental Words' we have created a framework to support the participants in an innovative way, while also introducing them to different networks to aid them in developing their professional practice. By revealing to each set of participants the language of their collaborators, we have shown them at first-hand the limits of their own worlds, while demonstrating the potential for embracing others. Moving away from a practice of teaching science and poetry separately is not easy, but in demonstrating to people from an early age how the two disciplines can work in unison we can better equip a future generation of scientists and poets to at least consider how they might more effectively work together.

In the Introduction I also discussed how, just as poetry can be characterised using a scientific approach, frontier science can also be analysed using modern forms of literary criticism. Similarly, poetry has been used as an effective facilitatory tool in the co-creation of science, for example by exploring how citizens are affected by environmental disasters,[5] and the effects of compassion fatigue in nurses.[6] My own research has likewise demonstrated how poetry can be used to better understand how different publics are affected by 'traditionally' scientific topics such as environmental change, and how poetry might be used as an effective tool through which to communicate scientific research to a wider audience.[7] In writing this book I wanted to present an aspirational account of how the two disciplines could work together, and in doing so hopefully inspire current and future generations of scientists and poets to see that their worlds are not mutually exclusive, but rather complementary in nature. I hope that I have gone some way to achieving this goal, with the scientists that feature in this book demonstrating that there is a congruence to be found between the two disciplines (although for the reasons that I have outlined above it is perhaps unreasonable to expect any 'expert' to self-identify as both a scientist and poet). To illustrate this point, I would like to return to the expert who inspired my own interest in this subject, and from whose poetry this book takes its title. Despite his reservations about the new technologies that accompanied the scientific achievements of his time, Poe had an interest in science, and like Byron before him, he was especially fascinated by astronomy and cosmology. This interest went beyond a merely superficial curiosity, and in his prose poem 'Eureka', Poe was arguably the first person to conceive and commit to writing the underlying concept of the Big Bang, that is, that our universe began by expanding outwards from a single point.[8] While not all of the claims that he made in 'Eureka' were true, Poe was able to outline a revolutionary theory of how the universe might have come into being, despite having no formal training or qualifications in science. However, notwithstanding these cosmological predictions, Poe himself was absolutely certain how he wanted both himself and 'Eureka' to be remembered, writing in the preface:

> *What I here propound is true:* —therefore it cannot die: —or if by any means it be now trodden down so that it die, it will "rise again to the Life Everlasting."

> Nevertheless it is as a Poem only that I wish this work to be judged after I am dead.[9]

Perhaps choice is something that is simply built into our very nature and identity. If this is indeed the case, then I hope that this book has at least demonstrated what can be achieved if we allow ourselves to first accept and then explore beyond the limits of our own codified languages. Poetry and science are both trying to make sense of the world in which we live, striving to describe nature in terms so absolute that they cannot possibly be denied, and thereby ensuring the posterity of both the author and the wider discipline. The irony, of course, is that neither the poet nor the scientist will prevail over nature itself. And yet, as both of them search in vain for a chance of immortality in an unforgiving and forgetful world, they could do far worse than to remember the words of the German polymath, Johann von Goethe:

> Science arose from poetry, and . . . when times change the two can meet again on a higher level as friends.[10]

Notes

Notes to the Introduction

1 E. A. Poe, *The Complete Tales & Poems of Edgar Allan Poe* (New York: Race Point Publishing, 2014), p. 827.

2 J. Keats, *The Complete Poems*, ed. John Barnard (Harmondsworth: Penguin, 2nd edn, 1977), p. 431.

3 www.dansimpsonpoet.co.uk (accessed 9 August 2018).

4 B. Williams, 'Wittgenstein and idealism', *Royal Institute of Philosophy Supplements*, 7 (1973), 76–95.

5 Aristotle, *Poetics*, trans. Anthony Kenny (Oxford: Oxford University Press, 2013).

6 See, for example, D. Tucan, 'The quarrel between poetry and philosophy: Plato – a sceptical view on "poetry"', *Procedia-Social and Behavioral Sciences*, 71 (2013), 168–75.

7 Aristotle, *Physics*, trans. R. Waterfield, ed. D. Bostock (Oxford: Oxford University Press, 2008); Aristotle, *The History of Animals*, trans. D. W. Thompson (Oxford: Clarendon Press, 1910).

8 It is also obviously the case that more modern poetic forms such as the elegy (Ovid, *c.* AD 1) and the clerihew (Bentley, *c.* 1900) are absent from Aristotle's 'data'.

9 See, for example, A. Brink, 'Transgressions: a quantum approach to literary deconstruction', *Journal of Literary Studies*, 1.3 (1985), 10–26; S. Reinharz, 'Friends or foes: gerontological and feminist theory', in Marilyn Pearsall (ed.), *The Other within Us: Feminist Explorations of Women and Aging* (Abingdon: Routledge, 2018), pp. 73–94.

10 R. B. Bradbury, A. Kyrkos, A. J. Morris, S. C. Clark, A. J. Perkins, and J. D. Wilson, 'Habitat associations and breeding success of yellowhammers on lowland farmland', *Journal of Applied Ecology*, 37.5 (2000), 789–805.

11 J. Clare, *Major Works*, ed. E. Robinson and D. Powell (Oxford: Oxford University Press, 2008), p. 230.

12 See, for example, C. Oldham, 'Natural history and folk-lore', *Nature*, 124 (1929), 229.

13 M. Midgley, *Science and Poetry* (Abingdon: Routledge, 2006).

14 J. Holmes, *Science in Modern Poetry: New Directions* (Liverpool: Liverpool University Press, 2012).

15 P. Middleton, *Physics Envy: American Poetry and Science in the Cold War and After* (Chicago: University of Chicago Press, 2015).

16 D. Brown, *The Poetry of Victorian Scientists: Style, Science and Nonsense* (Cambridge: Cambridge University Press, 2013).

17 R. Crawford (ed.), *Contemporary Poetry and Contemporary Science* (Oxford: Oxford University Press, 2006). See, for example, G. Beer, 'Translation or transformation? The relations of literature and science', *Notes and Records of the Royal Society of London*, 44.1 (1990), 81–99; G. Beer, *Darwin's Plots: Evolutionary Narrative in Darwin, George Eliot and Nineteenth-Century Fiction* (Cambridge: Cambridge University Press, 2009).

18 One notable exception being an essay by Miroslav Holub, who is the subject of one of the chapters of this book.

19 See, for example, L. Tsui, 'Effective strategies to increase diversity in STEM fields: a review of the research literature', *The Journal of Negro Education*, 76.4 (2007), 555–81.

20 O. Classe, *Encyclopedia of Literary Translation into English: A–L* (Abingdon: Routledge, 2000).

21 M. Priestman, *The Poetry of Erasmus Darwin: Enlightened Spaces, Romantic Times* (Abingdon: Routledge, 2016).

Notes to Chapter 1

1 J. Davy, *Memoirs of the Life of Sir Humphry Davy* (1836), 2 vols (Cambridge: Cambridge University Press, 2011), I, p. 36.

2 An educational institute that teaches children theology, normally to prepare them for a life working in the Roman Catholic Church.

3 R. Lamont-Brown, *Humphry Davy* (Stroud: The History Press, 2004), p. 11.

4 J. A. Paris, *The Life of Sir Humphry Davy* (1831), 2 vols (Cambridge: Cambridge University Press, 2011), I, p. 12.

5 Unless stated otherwise, all transcriptions of Davy's letters are from the excellent *Davy Letters Project*, the online version of which can be found at www. davy-letters.org.uk (accessed 15 February 2019). Since this book was written, this database has been superseded by *The Collected Letters of Sir Humphry Davy*, ed. T. Fulford and S. Ruston, 4 vols (Oxford: Oxford University Press, 2018).

6 J. Z. Fullmer, *Young Humphry Davy: The Making of an Experimental Chemist* (Philadelphia, PA: American Philosophical Society, 2000), p. 22.

7 This translation is provided by the physicist, educator, and poet Jean-Patrick Connerade, who says of it: 'I made the sky infinite in my translation, because *l'Etre Infini* in French is more divine than the *Infinite Being* would sound in English' (personal communication, 2018). While this non-author-sanctioned translation is at odds with the criteria I set out in the Introduction to this book, it has been included here out of interest to the reader.

8 Davy, *Memoirs of the Life of Sir Humphry Davy*, I, p. 21.

9 *Ibid.*, p. 22.

10 Davy, *Memoirs of the Life of Sir Humphry Davy*, I, p. 34.

11 See, for example, M. Hindle, 'Nature, power, and the light of suns: the poetry of Humphry Davy', *The Charles Lamb Bulletin*, 157 (2013), 38–54.

12 Fullmer, *Young Humphry Davy*, p. 153.

13 The *Annual Anthology* was effectively set up as a means through which to publish the work of Southey, Coleridge, and their friends, though mainly the work of Southey and Coleridge.

14 S. Ruston, 'When respiring gas inspired poetry', *The Lancet*, 381 (2013), 366–7 (p. 367).

15 See, for example, M. H. Abrams, *The Milk of Paradise: The Effect of Opium Visions on the Works of DeQuincey, Crabbe, Francis Thompson, and Coleridge* (New York: Octagon Books, 1971).

16 W. Wordsworth and D. Wordsworth, *The Early Letters of William and Dorothy Wordsworth (1787–1805)*, ed. E. de Sélincourt (Oxford: Clarendon Press, 1935), p. 244.

17 Davy's adopted grandparent, Tonkin.

18 Fullmer, *Young Humphry Davy*, p. 137.

19 Paris, *The Life of Sir Humphry Davy*, I, p. 40.

20 Davy probably learned about Spinozism from Coleridge, who, along with some of the other Romantic poets, saw in Spinozism a religion of nature.

21 Davy, *Memoirs of the Life of Sir Humphry Davy*, I, p. 390.

22 *Ibid.*, p. 56.

23 Davy is referring to the poem 'Extract from an Unfinished Poem on Mount's Bay', published in the *Annual Anthology* in 1799.

24 T. Beddoes and T. Longman, *Contributions to Physical and Medical Knowledge, Principally from the West of England* (Bristol: Biggs & Cottle, 1799).

25 Roget would go on to publish the *Thesaurus of English Words and Phrases*, and also conducted much research into the use of nitrous oxide as an anaesthetic.

26 Benjamin Thompson was an American-born British physicist and inventor who

was later made a Count of the Holy Roman Empire following his work as a personal assistant and advisor to the prince-elector, Charles Theodore.

27 W. Wordsworth and S. T. Coleridge, *Lyrical Ballads 1798 and 1802*, ed. F. Stafford (Oxford: Oxford University Press, 2013), p. 107.

28 A full copy of this poem can be found at http://digital.library.upenn.edu/women/barbauld/1811/1811.html (accessed 15 February 2019).

29 A. Johns-Putra, '"Blending science with literature": the Royal Institution, Eleanor Anne Porden and *The Veils*', *Nineteenth-Century Contexts*, 33.1 (2011), 35–52 (p. 38).

30 H. Davy, 'Select modern poetry', *The Gentleman's Magazine*, 76 (1806), 1148.

31 Davy composed this on the opening night of *The Honeymoon*, after a friend had informed him that the play was missing a prologue.

32 S. Ruston, *Creating Romanticism: Case Studies in the Literature, Science and Medicine of the 1790s* (London: Palgrave Macmillan, 2013), p. 1.

33 H. Davy, 'On the fire-damp of coal mines, and on methods of lighting the mines so as to prevent its explosion', *Philosophical Transactions of the Royal Society of London*, 106 (1816), 1–22.

34 D. Goldstein, D. Goldstein and B. Thompson, *Polarized Light, Revised and Expanded* (Boca Raton, FL: CRC Press, 2003), p. xx.

35 Davy uses the two spellings of Stephenson and Stevenson interchangeably.

36 A tree from Sri Lanka.

37 S. Ruston, 'Humphry Davy in 1816: letters and the lamp', *Wordsworth Circle*, 48 (2017), 6–16 (p. 13).

38 H. Davy, 'Some observations and experiments on the papyri found in the ruins of Herculaneum', *Philosophical Transactions of the Royal Society of London*, 111 (1821), 191–208.

39 Spelling is correct in this edition.

40 Lord Byron, *The Major Works*, ed. J. J. McGann (Oxford: Oxford University Press, 2008), p. 411.

41 In his biography of his brother, John Davy notes that this poem was composed during the Great Storm of November 1824 (a hurricane-force wind that affected the south coast of England), and that the poem should have been written during this storm, and highlighted as such by Davy in the title, is perhaps significant given that Byron himself died during a thunderstorm.

42 A form of crudely processed oil that was used in old lamps.

43 Davy, *Memoirs of the Life of Sir Humphry Davy*, II, p. 169.

44 The same nepotistic approach that Davy himself had benefited from in his battles with Stephenson.

45 Banks first made his name on the 1766 natural history expedition to Newfoundland

and Labrador, and after also taking part in Captain James Cook's first great voyage around the world from 1768 to 1771 he was very well connected in the Admiralty.

46 In 1816 Davies Giddy changed his surname to Gilbert, in order that he and his wife could inherit her uncle's estate.

47 Faraday was eventually elected to the Royal Society in 1824.

48 Davy, *Memoirs of the Life of Sir Humphry Davy*, II, p. 157.

49 R. Holmes, *The Age of Wonder: How the Romantic Generation Discovered the Beauty and Terror of Science* (New York: Vintage, 2010).

50 This failure resulted in strained relationships between the British government and the Royal Society and was a contributing factor to the dissolution in 1828 of the Board of Longitude (of which Davy had been the chairman). The Board of Longitude was at that time the only body that funded scientific endeavours using state money; after the board was dissolved it was replaced by the Resident Committee for Scientific Advice for the Admiralty, one of whose scientific advisors was Michael Faraday.

51 In 1827 Davies Gilbert was elected as Davy's successor.

52 H. Davy, *The Collected Works: Salmonia, or Days of Fly-Fishing: In a Series of Conversations; with Some Account of the Habits of Fishes Belonging to the Genus Salmo* (London: John Murray, 1828); H. Davy, *Consolations in Travel or the Last Days of a Philosopher* (London: John Murray, 1830).

53 See, for example, C. Flaherty, 'A recently rediscovered unpublished manuscript: the influence of Sir Humphry Davy on Anne Brontë', *Brontë Studies*, 38.1 (2013), 30–41.

54 Davy, *Consolations in Travel or the Last Days of a Philosopher*, p. ix.

55 *Ibid.*, p. 252.

56 Wahida Amin, 'The Poetry and Science of Humphry Davy', PhD thesis, University of Salford, 2013, p. 330.

57 C. J. Thoman, 'Sir Humphry Davy and Frankenstein', *Journal of Chemical Education*, 75.4 (1998), 495.

58 E. C. Bentley, *The Complete Clerihews* (Cornwall: House of Stratus, 2008), p. 38. In some later versions of this poem, 'Detested' is changed to 'Abominated'.

Notes to Chapter 2

1 Wentworth Bequest, Vol. CCLXXIII, Family correspondence, etc.; 1833–1852. Contents: 1. Letters from Hester King, daughter of Peter, 7th Baron Lovelace, and (1843) wife of the Rev. George William Craufurd, 3rd Bart., to Louisa, wife of Robert R. Noel, British Library.

2 It was Lord Byron who decided on the name Ada for his daughter.

3 *Lord Byron: Selected Letters and Journals*, ed. L. A. Marchand (Cambridge, MA: Belknap Press of Harvard University Press, 1982), p. 65.

4 Lord Byron, *The Major Works*, ed. J. J. McGann (Oxford: Oxford University Press, 2008), p. 139.

5 P. Douglass, 'The madness of writing: Lady Caroline Lamb's Byronic identity', *Pacific Coast Philology*, 34.1 (1999), 53–71 (p. 53).

6 Byron, *The Major Works*, p. 381. Dimity is a lightweight cotton fabric.

7 A. A. Lovelace, 'Sketch of the Analytical Engine invented by Charles Babbage, Esq. By L. F. Menabrea, of Turin, officer of the military engineers', *Scientific Memoirs*, 3 (1843), 666–731 (p. 731).

8 B. A. Toole, *Ada, the Enchantress of Numbers: A Selection from the Letters of Lord Byron's Daughter and Her Description of the First Computer* (Sausalito, CA: Strawberry Press, 1992).

9 *Ibid.*, p. 198.

10 C. Babbage, *Passages from the Life of a Philosopher* (1864) (Cambridge: Cambridge University Press, 2011), p. 136.

11 Lovelace, 'Sketch of the Analytical Engine invented by Charles Babbage', pp. 694, 722.

12 A. M. Turing, 'Computing machinery and intelligence', *Mind*, 59.236 (1950), 433–60 (p. 450).

13 A. M. Turing, 'On computable numbers, with an application to the Entscheidungsproblem', *Proceedings of the London Mathematical Society*, series 2, 42.1 (1937), 230–65.

14 Toole, *Ada, the Enchantress of Numbers*, p. 218.

15 Wentworth Bequest, Vol. CCLXXIII, Family correspondence, etc.; 1833–1852. Contents: 1. Letters from Hester King, daughter of Peter, 7th Baron Lovelace, and (1843) wife of the Rev. George William Craufurd, 3rd Bart., to Louisa, wife of Robert R. Noel, British Library.

16 *Ibid.*

17 *Ibid.*

18 Toole, *Ada, the Enchantress of Numbers*, p. 264.

19 E. B. Escott, 'Gleanings far and near', *The Mathematical Gazette*, 13.186 (1927), 270.

20 The correspondence is available in the British Library in the Western Manuscripts Collection, ADD MS 37190–ADD MS 37194.

21 Babbage, *Passages from the Life of a Philosopher*, p. 112.

22 Byron, *The Major Works*, p. 104.

23 *Ibid.*, p. 139.

24 Lovelace underlined a lot of words and phrases in her letters. Here I use italics in place of underlining.

25 Toole, *Ada, the Enchantress of Numbers*, p. 138.

26 I.e. mathematics, optics, astronomy, and physics.

27 Toole, *Ada, the Enchantress of Numbers*, p. 136.

28 R. P. Feynman, R. B. Leighton, and M. Sands, *The Feynman Lectures on Physics, Desktop Edition Volume I* (New York: Basic Books, 2013), p. 3.

29 Toole, *Ada, the Enchantress of Numbers*, p. 179.

30 Despite much investigation, this interpretation could not be found in any of Lovelace's correspondence.

31 Toole, *Ada, the Enchantress of Numbers*, p. 215.

32 Anon., 'Stanzas by the Late Lady Lovelace on Miss Nightingale (from the *London Morning Advertiser*)', *Hobart Town Daily Courier*, 9 June 1856, p. 3.

33 Byron, *The Major Works*, p. 258.

34 Anon., 'Stanzas by the Late Lady Lovelace on Miss Nightingale', p. 3.

35 *Ibid.*

36 With the death in 1798 of his great-uncle, the fifth Lord Byron, George became the sixth Baron Byron of Rochdale at the age of 10.

37 R. C. Dallas, *Recollections of the Life of Lord Byron: From the Year 1808 to the End of 1814 ... Including Various Unpublished Passages of His Works* (London: C. Knight, 1824), p. 207.

38 See, for example, M. Davis and V. Davis, 'Mistaken ancestry: the Jacquard and the computer', *Textile*, 3.1 (2005), 76–87.

39 Lord Byron, *The Complete Poetical Works* (Boston: Houghton Mifflin, 1905), p. xxi.

40 Wentworth Bequest, Vol. CCLXXIII, Family correspondence, etc.; 1833–1852. Contents: 1. Letters from Hester King, daughter of Peter, 7th Baron Lovelace, and (1843) wife of the Rev. George William Craufurd, 3rd Bart., to Louisa, wife of Robert R. Noel, British Library.

41 Toole, *Ada, the Enchantress of Numbers*, p. 394.

42 *Ibid.*, p. 394.

43 *Ibid.*, p. 54.

44 Byron, *The Major Works*, p. 272.

45 See, for example, R. B. Stothers, 'The great Tambora eruption in 1815 and its aftermath', *Science*, 224.4654 (1984), 1191–8.

46 Wentworth Bequest, Vol. CCLXXIII, Family correspondence, etc.; 1833–1852. Contents: 1. Letters from Hester King, daughter of Peter, 7th Baron Lovelace, and (1843) wife of the Rev. George William Craufurd, 3rd Bart., to Louisa, wife of Robert R. Noel, British Library.

47 Toole, *Ada, the Enchantress of Numbers*, p. 398.

48 Every modern computer can be considered a realisation of the universal computer.

49 Toole, *Ada, the Enchantress of Numbers*, p. 203.

Notes to Chapter 3

1 L. Campbell and W. Garnett, *The Life of James Clerk Maxwell* (London: Macmillan, 1882), p. 638.

2 Whereupon he added the surname 'Maxwell' to his previous name 'John Clerk'.

3 Psalm 119 is an example of an acrostic alphabet poem, divided into twenty-two stanzas for each of the twenty-two letters of the Hebrew alphabet. The eight couplets that make up each stanza each start with the corresponding letter of the alphabet – a pattern that hardly makes the task of memorising all twenty-two stanzas any easier.

4 M. McCartney, 'The poetic life of James Clerk Maxwell', *BSHM Bulletin*, 26.1 (2011), 29–43 (p. 32).

5 See, for example, R. Flood, M. McCartney, and A. Whitaker (eds), *James Clerk Maxwell: Perspectives on his Life and Work* (Oxford: Oxford University Press, 2014).

6 Campbell and Garnett, *The Life of James Clerk Maxwell*. All of Maxwell's poetry that appears in this chapter is taken from this biography, which splits the forty-three listed poems into three sections: 'Juvenile verse and translations', 'Occasional pieces', and 'Serio-comic verse'. The first of these sections concerns Maxwell's writing up until he left the Edinburgh Academy, while the second and third appear to have been roughly split depending on whether they concerned matters of science (Serio-comic) or not (Occasional), although there are potentially some exceptions.

7 J. C. Maxwell, *The Scientific Letters and Papers of James Clerk Maxwell*, ed. P. M. Harman, 3 vols (Cambridge: Cambridge University Press, 2002).

8 Campbell and Garnett, *The Life of James Clerk Maxwell*, p. 583.

9 *Ibid.*, p. 64.

10 *Ibid.*, p. 69.

11 The norm at the time was for research to be first presented to a society or learned body. If it met approval at this stage it would then receive an additional round of peer review by a selected member or members from the organisation before being published in its transactions and/or proceedings.

12 Both poems were originally untitled. The titles that appear here are those that were given by Campbell when the poems were first published in his biography of Maxwell.

13 Campbell and Garnett, *The Life of James Clerk Maxwell*, p. 590.

14 Maxwell was in the school year directly below Tait.

15 Campbell and Garnett, *The Life of James Clerk Maxwell*, p. 86.

16 J. C. Maxwell, 'XXXV. – On the Theory of Rolling Curves', *Earth and*

Environmental Science Transactions of the Royal Society of Edinburgh, 16.5 (1849), 519–40; J. C. Maxwell, 'IV.—On the equilibrium of elastic solids', *Earth and Environmental Science Transactions of the Royal Society of Edinburgh*, 20.1 (1853), 87–120.

17 Note that while the dates of these two papers are given as 1849 and 1853, they were initially presented to the Royal Society of Edinburgh in 1849 and 1850, respectively.

18 As with Davy (see Chapter 1), some of Maxwell's earliest poems were translations.

19 Campbell and Garnett, *The Life of James Clerk Maxwell*, p. 612.

20 C. M. Neale, *The Senior Wranglers of the University of Cambridge, from 1748 to 1907. With Biographical, &c., Notes* (Bury St Edmunds: F.T. Groom and Son, 1907), p. 33.

21 Campbell and Garnett, *The Life of James Clerk Maxwell*, p. 623.

22 A. Warwick, *Masters of Theory: Cambridge and the Rise of Mathematical Physics* (Chicago: University of Chicago Press, 2003), p. 25.

23 J. C. Maxwell, 'On the transformation of surfaces by bending', *Transactions of the Cambridge Philosophical Society*, 9.4 (1854), 445.

24 J. C. Maxwell, 'XVIII. – Experiments on Colour, as perceived by the Eye, with Remarks on Colour-Blindness', *Transactions of the Royal Society of Edinburgh*, 21.2 (1857), 275–98.

25 Campbell and Garnett, *The Life of James Clerk Maxwell*, p. 248.

26 *Ibid.*, p. 600. Maxwell presented this poem to Campbell when they met shortly after the death of his father.

27 B. Mahon, *The Man who Changed Everything: The Life of James Clerk Maxwell* (Hoboken, NJ: John Wiley & Sons, 2004), p. 7.

28 J. C. Maxwell, *On the Stability of the Motion of Saturn's Rings* (Cambridge: Macmillan, 1859).

29 See, for example, J. S. Reid, 'Maxwell at Aberdeen', in R. Flood, M. McCartney, and A. Whitaker (eds), *James Clerk Maxwell: Perspectives on his Life and Work* (Oxford: Oxford University Press, 2014), p. 38.

30 Maxwell was the best man at Campbell's own wedding in May 1858.

31 Campbell and Garnett, *The Life of James Clerk Maxwell*, p. 276.

32 *Ibid.*, p. 609.

33 McCartney, 'The poetic life of James Clerk Maxwell', p. 39.

34 The same award had been bestowed on Humphry Davy for his work on the Davy lamp (see Chapter 1).

35 J. C. Maxwell, 'Xxv. on physical lines of force: Part i.–the theory of molecular vortices applied to magnetic phenomena', *The London, Edinburgh, and Dublin Philosophical Magazine and Journal of Science*, 21.139 (1861), 161–75. Maxwell

himself had favoured a notation that made use of quaternions, a number system that extends the complex numbers.

36   J. C. Maxwell, 'VIII. A Dynamical Theory of the Electromagnetic Field', *Philosophical Transactions of the Royal Society of London*, 155 (1865), 459–512 (p. 499).

37   C. C. Gaither and A. E. Cavazos-Gaither, *Physically Speaking: A Dictionary of Quotations on Physics and Astronomy* (Boca Raton, FL: CRC Press, 1997), p. 132.

38   F. J. C. Hearnshaw, *The Centenary History of King's College, London, 1828–1928* (London: G. G. Harrap, 1929), p. 248.

39   See, for example, H. Lidbetter, 'Henry Cavendish and Asperger's syndrome: a new understanding of the scientist', *Personality and Individual Differences*, 46.8 (2009), 784–93.

40   R. Sviedrys, 'The rise of physics laboratories in Britain', *Historical Studies in the Physical Sciences*, 7 (1976), 405–36 (p. 427).

41   D.-W. Kim, *Leadership and Creativity: A History of the Cavendish Laboratory, 1871–1919* (Dordrecht: Springer Science & Business Media, 2002), p. 14.

42   See M. Longair, *Maxwell's Enduring Legacy: A Scientific History of the Cavendish Laboratory* (Cambridge: Cambridge University Press, 2016), for a complete description of Maxwell's influence on the Cavendish Laboratory.

43   P. Theerman, 'James Clerk Maxwell and religion', *American Journal of Physics*, 54.4 (1986), 312–17 (p. 312).

44   Campbell and Garnett, *The Life of James Clerk Maxwell*, p. 593.

45   I. H. Hutchinson, 'The genius and faith of Faraday and Maxwell', *The New Atlantis*, 41 (2014), 81–99 (p. 97).

46   J. C. Maxwell, 'Molecules', *Nature*, 8.204 (1873), 437–41 (p. 441).

47   Maxwell was an active member of the BAAS, giving several lectures and serving as president of Section A (mathematics and physics) for the 1870 meeting in Liverpool (Campbell and Garnett, *The Life of James Clerk Maxwell*, p. 71). Furthermore, he also led the work of the BAAS 1863 report which defined most of the electrical units in use today; R. Glazebrook, *James Clerk Maxwell and Modern Physics* (London: Cassell, 1896).

48   Campbell and Garnett, *The Life of James Clerk Maxwell*, p. 640.

49   Maxwell, 'Molecules', p. 440.

50   B. Stewart and P. G. Tait, *The Unseen Universe: Or, Physical Speculations on a Future State* (New York: Macmillan, 1875).

51   D. Brown, *The Poetry of Victorian Scientists: Style, Science and Nonsense* (Cambridge: Cambridge University Press, 2013), p. 91.

52   Campbell and Garnett, *The Life of James Clerk Maxwell*, p. 638.

53   Entozoa are animal parasites, such as tapeworms, that live within the body of a host.

54 P. Collins, 'The BA at play', *New Scientist*, 91 (1981), 551–2 (p. 552).

55 This is not to be confused with another, earlier poem, written by Maxwell in 1873, which is also entitled 'Molecular Evolution', and which can be read alongside his 1873 lecture to the British Association as a further comment on his views against the materialism that was being championed by Tyndall et al.

56 See, for example, S. Pratt-Smith, 'Boundaries of perception: James Clerk Maxwell's poetry of self, senses and science', in R. Flood, M. McCartney, and A. Whitaker (eds), *James Clerk Maxwell: Perspectives on his Life and Work* (Oxford: Oxford University Press, 2014), p. 234.

57 At the same age, and from the same causes, as his mother.

58 J. C. Maxwell, *The Scientific Papers of James Clerk Maxwell*, ed. W. D. Niven (Cambridge: Cambridge University Press, 1890), p. xii.

59 Campbell and Garnett, *The Life of James Clerk Maxwell*, p. ix.

60 P. G. Tait, 'James Clerk Maxwell obituary', *Proceedings of the Royal Society of Edinburgh*, 10 (1878), 331–9 (p. 338).

Notes to Chapter 4

1 R. Ross, *Memoirs: With a Full Account of the Great Malaria Problem and its Solution* (London: John Murray, 1923), p. 226.

2 A leave of absence from duty granted to members of the army.

3 Ross's great-great-grandfather had been a director of the British East India Company.

4 Ross, *Memoirs*, p. 20.

5 *Ibid.*, p. 21.

6 R. Ross, *In Exile* (Liverpool: Philip, Son & Nephew, privately printed, 1906), p. 61. This was later republished in various guises, including as part of the collection presented in *Philosophies*. However, these later versions do not include the stanzas dedicated to his aunt.

7 An early form of private elementary school, which predated the introduction of compulsory education in England in 1880. These schools were usually run by women from their homes.

8 Ross, *Memoirs*, p. 21.

9 R. Ross, *Fables and Satires* (London: Harrison & Sons, 1928).

10 *Ibid.*, p. 65.

11 *Ibid.*, p. 64.

12 R. L. Mégroz, *Ronald Ross, Discoverer and Creator* (London: Allen & Unwin, 1931), p. 137.

13 Ross, *Memoirs*, p. 25.

14 *Ibid.*, p. 24. This is likely to be the soldier carrying a torch who appears on the left fresco of Raphael's *Liberation of Saint Peter*, located in the Vatican.

15 Lord Byron, *The Major Works*, ed. J. J. McGann (Oxford: Oxford University Press, 2008), p. 276.

16 Ross, *Memoirs*, p. 28.

17 Ross's father was, apparently, extremely reluctant to return to India in 1875 for his final stint of military service there.

18 Frozen or frosty.

19 Mégroz, *Ronald Ross, Discoverer and Creator*, p. 36.

20 R. Ross, *The Spirit of Storm: A Romance* (London: Methuen, 1896).

21 J. W. von Goethe, *The Poems of Goethe: Translated in the Original Metres: By Edgar Alfred Bowring* (London: John W. Parker and Son, 1853), p. 147.

22 Ross, *Memoirs*, p. 36.

23 A copy of this sketch can be found in M. Kemp, 'A feverish imagination', *Nature*, 451 (2008), 1056, https://www.nature.com/articles/4511056a (accessed 23 January 2019).

24 Ross, *Memoirs*, p. 8. Many years later, while at a conference in St Louis in the United States, Ross was approached by a distinguished American chemist, who confirmed that his uncle had indeed made several significant advances in the field.

25 This was despite the assumption at the time that the LSA was one of the easier qualifications that doctors could pass in order to become medical professionals. Keats also qualified in medicine via this route.

26 Now Mumbai.

27 Now Chennai.

28 Ross, *Memoirs*, p. 43.

29 *Ibid.*, p. 44.

30 *Ibid.*

31 The British Empire was largely to blame for the economic and physical ill health of these 'swarming millions'.

32 Both of these poems would later appear in the 1910 collection *Philosophies*, as 'Preludes' alongside 'India', in the 'In Exile' sequence.

33 Marsh.

34 R. Ross, *Philosophies* (London: John Murray, 1910), p. 2.

35 Lantern.

36 Grotesque or bizarre.

37 Ross, *Philosophies*, p. 2.

38 O. M. Mitchell, *The Orbs of Heaven* (London: Routledge, Warne and Routledge, 1858).

39 Ross, *Memoirs*, p. 50.

40 R. Ross, *Edgar, or the New Pygmalion; and the Judgement of Tithonus* (Madras: Higginbotham, 1883).

41 Ross, *Memoirs*, p. 76.

42 *Ibid.*

43 In his autobiography, Ross is quick to point out that the distractions of his wedding were to blame for his poor performance in the diploma. He also suggests that the only reason he passed was because his examiners were impressed by his use of the binomial theorem to solve a question on population.

44 See, for example, B. Atalić and S. Fatović-Ferenčić, 'Emanuel Edward Klein – the father of British microbiology and the case of the animal vivisection controversy of 1875', *Toxicologic Pathology*, 37.6 (2009), 708–13.

45 R. Ross, *The Child of Ocean: A Romance* (London: Remington, 1889).

46 R. Ross, *Lyra Modulata* (Liverpool: C. Tinling, privately printed, 1911).

47 Now Myanmar.

48 Now Pakokku.

49 Now Bengaluru.

50 R. Ross, *The Deformed Transformed* (Bangalore: Spectator Press, privately printed, 1890).

51 Ross, *Memoirs*, p. 97.

52 Ross, *Philosophies*, p. 49.

53 Ross, *Memoirs*, p. 98.

54 This parasite is responsible for the deadliest form of the disease.

55 Malaria can also be transmitted through the sharing of needles and blood transfusions, although this is much less common.

56 The reason for these recurrences is probably because Ross's father was treated with quinine, which while effective at treating the fever and other related symptoms of malaria, does not kill the parasites.

57 R. Ross, 'Some objections to haematozoic theories of malaria', *The Medical Reporter*, 2 (1893), 65–71.

58 A type of single-celled microorganism.

59 In 1907 Laveran would win the Nobel Prize in Physiology or Medicine for his work in determining that parasitic protozoans were responsible for infectious diseases such as malaria.

60 Ross, *Philosophies*, p. 21.

61 Following this meeting, Ross devised a high-quality, portable telescope, which he had built and took back to India with him.

62 R. Ross, *The Revels of Orsera: A Mediaeval Romance* (London: John Murray, 1920).

63 As well as the *Child of Ocean* debacle, Ross was also dismayed at one publisher's suggestion of changing the ending of *The Spirit of Storm*.

64   Awarded by the Army Medical Service at Netley hospital, this award also came with a purse of 75 guineas. This was the equivalent of approximately £33,000 in 2019, calculated according to the labour cost of government expenditure via the website www.measuringworth.com (accessed 15 February 2019).

65   A slender whip-like appendage found on microorganisms; it was the lashing of these flagella that Laveran first noticed in a blood sample from a dead Tunisian soldier, causing him to properly observe the *Plasmodium* parasite and develop his hypothesis that it was responsible for the malaria fever.

66   W. F. Bynum and C. Overy (eds), *The Beast in the Mosquito: The Correspondence of Ronald Ross and Patrick Manson* (Amsterdam and Atlanta, GA: Rodopi, 1998), p. 136.

67   With typical humility, Ross later declared that this date should be named 'World Mosquito Day', in celebration of his discovery.

68   Ross, *Memoirs*, p. 226.

69   R. Ross, 'Report on the Cultivation of Proteosoma Labbé, in Grey Mosquitos', *The Indian Medical Gazette*, 33.12 (1898), 448–51 (p. 451).

70   Ross, *Memoirs*, p. 269.

71   Also known as visceral leishmaniasis, this is another deadly parasitic disease, which if untreated has an almost 100 per cent mortality rate.

72   A vector is an organism that transmits a disease or parasite from one plant or animal to another.

73   World Health Organization, Global Health Observatory data repository, 2016.

74   M. P. Scott, 'Developmental genomics of the most dangerous animal', *Proceedings of the National Academy of Sciences of the United States of America*, 104.29 (2007), 11865–6 (p. 11865).

75   R. Ross, *The Prevention of Malaria* (London: John Murray, 1910).

76   R. Ross, *The Algebra of Space: Being a Brief Description of a System of Geometrical Algebra Placed on an Arithmetical Basis* (Liverpool: George Philip & Son, 1901).

77   The study and analysis of patterns of disease within defined populations.

78   The citation can be read in full at www.nobelprize.org/nobel_prizes/medicine/laureates/1902/ross-lecture.html (accessed 15 February 2019).

79   Neglecting any health and safety concerns, Grassi ate roundworm eggs from a heavily infected human corpse and upon finding fresh eggs in his own faeces, demonstrated that roundworm is transmitted through directly ingesting a contaminated source. In other words, Grassi really *did* 'swallow a live entozöon' in the name of science (see Chapter 3).

80   G. Cook, *Tropical Medicine: An Illustrated History of the Pioneers* (Cambridge, MA: Academic Press, 2007), p. 96.

81   R. Jones, *Mosquito* (London: Reaktion, 2012), p. 93.

82 It is interesting to draw parallels between Ross's opinion of Grassi and Davy's opinion of Stephenson, especially in their choice of defamatory insults (see Chapter 1).

83 R. Ross, Nobel Lecture, 12 December 1902, in *Nobel Lectures, Physiology or Medicine, 1901–1921* (Amsterdam: Elsevier, 1967), p. 105.

84 G. B. Grassi, *Studio di uno zoologo sulla malaria* (Rome: R. Accademia dei lincei, 1900).

85 *Ibid.*, p. 8.

86 Ross, *Memoirs*, p. 493.

87 Ross, *Poems*, p. 25.

88 Royal Institution of Great Britain, *Notices of the Proceedings at the Meetings of the Members of the Royal Institution of Great Britain: With Abstracts of the Discourses Delivered at the Evening Meetings 1920–1922* (London: William Clowes and Sons, 1924), p. 225.

89 *Ibid.*, p. 226.

90 A. C. Doyle, *The Poems of Arthur Conan Doyle – Collected Edition* (London: John Murray, 1922), p. 216. 'Boche' is a derisive term used to refer to someone from Germany, especially a German soldier in the First World War.

91 The Ross Institute was later moved to the London School of Hygiene and Tropical Medicine's Bloomsbury site, and in 1934 it was formally incorporated into the School.

92 Ross had fastidiously maintained almost all of his correspondence and other materials, amassing a large personal archive. A lot of these materials, including photographs, notebooks, press cuttings, and a sketch book, are now available via the Sir Ronald Ross collections at the London School of Hygiene and Tropical Medicine.

93 Ross, *Memoirs*. The autobiography won Ross the James Tait Black Memorial Prize when it was published.

94 E. Nye and M. Gibson, *Ronald Ross: Malariologist and Polymath: A Biography* (Dordrecht: Springer, 1997), p. 217.

95 The reason for the delay was that Masefield had first to write to the church to express his opinion that not enough had been done to commemorate Ross's life and work. During the service, as well as hearing Ross's poems read, the congregation also sang one of his hymnal compositions, 'Before Thy Feet I Fall'.

96 Anon., 'Lines for the centenary of Ronald Ross, scientist and poet', *The Times*, 13 June 1957.

97 This was the equivalent of approximately £12.6 million in 2019, calculated according to the labour cost of government expenditure via the website www.measuringworth.com (accessed 15 February 2019).

98 R. Ross, *Memories of Sir Patrick Manson* (London: Harrison & Sons, privately published, 1930). Manson had died in 1922, and so was unable to respond to these claims.

99 Royal Institution of Great Britain, *Notices of the Proceedings*, p. 227. This address, which was entitled 'Science and Poetry', took place on 4 June 1920, and also featured the recounting of Ross's torpedo escapade.

Notes to Chapter 5

1 M. Holub, *Poems Before and After* (Hexham: Bloodaxe, 2006), p. 309.

2 Now the Czech Republic.

3 M. Holub, *Supposed to Fly: A Sequence from Pilsen, Czechoslovakia*, trans. E. Osers (Hexham: Bloodaxe, 1996).

4 M. Holub, *Notes of a Clay Pigeon* (London: Secker and Warburg, 1977).

5 Holub, *Supposed to Fly*, p. 17.

6 A. Finlay, 'Interview: Miroslav Holub', *New Coin Poetry*, 32.2 (1996), 40–3 (p. 42).

7 Holub, *Supposed to Fly*, p. 33.

8 S. O'Shea, 'Interview with Miroslav Holub', *The Poetry Ireland Review* (autumn–winter 1990), p. 64.

9 Holub, *Poems Before and After*, p. 278.

10 Holub, *Supposed to Fly*, p. 73.

11 *Ibid.*, p. 72.

12 The two locations were separated by approximately 12 miles.

13 Holub, *Supposed to Fly*, p. 150.

14 These executions occurred on 17 November 1939, a date which is now known across the world as International Students' Day, in memory of the students who were executed.

15 Holub, *Supposed to Fly*, p. 85.

16 http://casopis.vesmir.cz/ (accessed 15 February 2019). He held this position until 1966 and remained on the editorial board of *Vesmír* until 1969.

17 M. Holub, 'Antibody production by lymphocytes after in vitro contact with bacterial antigen and transfer to new-born rabbits', *Nature*, 181.4602 (1958), 122. Lymphocytes are a type of white blood cell that forms part of the immune system.

18 This expression was originally derived from the title of the short novel 'Thaw', written by the Soviet writer Ilya Ehrenburg and first published in 1954.

19 I. B. Honeycutt, 'Interview with Czech poet Miroslav Holub', *The Virginia Quarterly Review* (2013), https://www.vqronline.org/interview/interview-czech-poet-miroslav-holub (accessed 15 February 2019).

20 J. Turnbull, T. Satterlee, and A. Raab, *The Global Game: Writers on Soccer* (Lincoln, NE: University of Nebraska Press, 2008), p. 283.

21 M. Holub, 'Rampage, or science in poetry', in R. Crawford (ed.), *Contemporary Poetry and Contemporary Science* (Oxford: Oxford University Press, 2006), p. 22.

22 Holub, *Poems Before and After*, p. 30.

23 William Carlos Williams (1883–1963) was a Puerto-Rican American poet who grew up in the United States, and trained and worked as a medical doctor for over forty years in the New Jersey town of Rutherford. Williams's writing was heavily influenced by the lives and experiences of his patients. He has been referred to by literary critics as the 'most important literary doctor since Chekhov'. R. Ellman and R. O'Clair (eds), *The Norton Anthology of Modern Poetry* (New York: W.W. Norton, 2nd edn, 1988), pp. 312–13.

24 I. Bamforth, 'Applied poetry', *Parnassus: Poetry in Review*, 25.1/2 (2001), 281–307 (p. 285).

25 P. Blažíček, 'Poetry of everyday life: the Květen group, 1956–1958 (young Czech writers)', *Ceska Literatura*, 49.1 (2001), 64–76 (p. 64).

26 M. Holub, 'Náš všední den je pevnina', *Květen* (September 1956), p. 551.

27 See, for example, G. Huang, *Whitmanism, Imagism, and Modernism in China and America* (Selinsgrove, PA: Susquehanna University Press, 1997).

28 Finlay, 'Interview: Miroslav Holub', p. 42.

29 M. Holub, *Selected Poems*, trans. Ian Milner and George Theiner (London: Penguin, 1967).

30 M. Holub, *Anděl na kolečkách* (Prague: Československý spisovatel., 1963); M. Holub, *Žít v New Yorku* (Prague: Melantrich, 1969).

31 J. Quinn, *Between Two Fires: Transnationalism and Cold War Poetry* (Oxford: Oxford University Press, 2015), p. 115.

32 S. Stepanchev, 'Metaphors and microscopes: talk with a Czech poet', *The New Leader*, 50.19 (1967), 13–14 (p. 13).

33 This would become known as the Brezhnev Doctrine.

34 The invasion involved five of the countries that had signed the collective defence treaty known as the Warsaw Pact: the Soviet Union, Bulgaria, Hungary, Poland, and East Germany. The invasion was officially referred to as 'Operation Danube'.

35 L. Vaculík, *Ludvík Vaculík: Two Thousand Words* [online], Prague Writer's Festival (2008), http://www.pwf.cz/rubriky/projects/1968/ludvik-vaculik-two-thousand-words_849.html (accessed 15 February 2019).

36 Although dismissed from his research institute, he was employed elsewhere as a scientist, and was not banned from practising as a doctor/scientist, as the KSČ did not ban medical doctors (Jan Čulík, personal communication, 2018).

37 Holub, *Poems Before and After*.
38 *Ibid.*, p. 64.
39 *Ibid.*, p. 102.
40 *Ibid.*, p. 161.
41 *Ibid.*, p. 220.
42 P. Bren, *The Greengrocer and his TV: The Culture of Communism after the 1968 Prague Spring* (Ithaca, NY: Cornell University Press, 2010), p. 58.
43 Holub would eventually marry three times, and remained with his third wife, Jitka, until his death. He also had three children.
44 P. Bridges, *Safirka: An American Envoy* (Kent, OH: Kent State University Press, 2000).
45 W. Wilde-Menozzi, 'Revising Miroslav Holub', *Southwest Review*, 88 (2003), 519–30; Jan Čulík, personal communication, 2018.
46 Holub, *Poems Before and After*, p. 214.
47 Finlay, 'Interview: Miroslav Holub', p. 42.
48 M. Holub, 'Angels of extermination, angels of exclusion', *Index on Censorship*, 26.5 (1997), 103–5 (p. 103).
49 William Carlos Williams, *The Collected Earlier Poems of William Carlos Williams* (Norfolk, CT: New Directions, 1951), p. 277.
50 Holub, *Poems Before and After*, pp. 375–6.
51 Now Corinth.
52 A. Camus, *The Myth of Sisyphus, and Other Essays* (New York: Vintage, 2018).
53 Holub, 'Rampage, or science in poetry', p. 24.
54 J. Limburg, *The Woman who Thought Too Much: A Memoir* (London: Atlantic Books, 2011), p. 170.
55 Holub, 'Rampage, or science in poetry', p. 13.
56 J. Holý and J. Čulík, 'Miroslav Holub', in J. Hardin, V. D. Mihailovich, and S. Serafin (eds), *Concise Dictionary of World Literary Biography: South Slavic and Eastern European Writers Vol. 4* (Detroit, MI: Cengage Gale, 2000), p. 141.
57 As noted earlier in this chapter, aside from *Supposed to Fly* there exists very little in terms of detailed biographical information about Holub's personal life.
58 Honeycutt, 'Interview with Czech poet Miroslav Holub'.
59 *Ibid.*
60 Holub, 'Rampage, or science in poetry', p. 24.
61 Finlay, 'Interview: Miroslav Holub', p. 40.
62 S. Heaney, 'The fully exposed poem', *Parnassus: Poetry in Review*, 5 (1983), 7. A sagittal section is a medical term for a cross-section obtained by slicing, actually or via imaging techniques, the body or any part of the body, in the sagittal plane, i.e. the plane that splits the body into left and right.

63 It should be noted that almost all of the English-language interviews that were conducted and recorded with Miroslav Holub were with literary journals, or publications with more of a poetic slant, rather than popular science magazines or their equivalent.

64 Wilde-Menozzi, 'Revising Miroslav Holub', p. 528.

65 *Ibid.*, p. 527.

66 R. M. Burke, 'Vanishing lungs: a case report of bullous emphysema', *Radiology*, 28.3 (1937), 367–71. Holub's poem is loosely dedicated to Burke. In print, the title is marked by an asterisk, which links to the name 'Burke' at the foot of the page.

67 Finlay, 'Interview: Miroslav Holub', p. 40.

68 Holub, *Poems Before and After*, p. 309.

69 Wilde-Menozzi, 'Revising Miroslav Holub', p. 527.

70 Finlay, 'Interview: Miroslav Holub', p. 40.

71 R. Kearney, *States of Mind: Dialogues with Contemporary Thinkers on the European Mind* (Manchester: Manchester University Press, 1995), p. 136.

72 *Ibid.*, p. 136.

73 M. Holub, *Shedding Life: Disease, Politics, and Other Human Conditions* (Minneapolis, MN: Milkweed Editions, 1997), p. 12.

74 This was a very serious matter, one that led to the StB threatening to charge Holub with spying for the Western powers (Jan Čulík, personal communication, 2018).

75 Honeycutt, 'Interview with Czech poet Miroslav Holub'. Prague 4 refers to the biggest municipality in Prague.

76 Holub, *Poems Before and After*, p. 440.

77 Honeycutt, 'Interview with Czech poet Miroslav Holub'.

78 Holub, *Poems Before and After*, p. 439.

79 A large apron-like fold of fatty tissue that hangs down from the stomach.

80 Established at the beginning of the Cold War to transmit uncensored news and information to audiences behind the Iron Curtain, Radio Free Europe ultimately played a significant role in the collapse of communism and the rise of democracies in post-communist Europe.

Notes to Chapter 6

1 R. Elson, *A Responsibility to Awe* (Manchester: Carcanet, 2001), p. 9.

2 *Ibid.*, pp. 149–59.

3 One such experiment that Elson remembered fondly was catching snowflakes on a black velvet cloth, before preserving them and examining them under a microscope.

4 Elson, *A Responsibility to Awe*, p. 151.

5 *Ibid.*, p. 28.

6 *Ibid.*, p. 150.

7 Smith College is a private and independent women's liberal arts college in Northampton, Massachusetts. Elson's father had been educated at Yale, where her mother (a New Englander) had worked as a librarian.

8 Elson, *A Responsibility to Awe*, p. 119.

9 At the time this telescope was operated by the Anglo-Australian Observatory (now the Australian Astronomical Observatory), which incidentally is where the author of this book spent the summer of his third year at university (MPhys in Physics with Space Science at Technology at the University of Leicester) observing irregularities in the dust of a spiral galaxy; this research was later published in the *Publications of the Astronomical Society of Australia*, where I was also listed as the second author.

10 T. G. Hawarden, R. A. W. Elson, A. J. Longmore, S. B. Tritton and H. G. Corwin, Jr, 'Early-type ("discless") galaxies with dust lanes', *Monthly Notices of the Royal Astronomical Society*, 196.4 (1981), 747–56.

11 Elson, *A Responsibility to Awe*, p. 154.

12 *Ibid.*, p. 24.

13 After an initial setback in terms of funding, the Hubble Space Telescope was due to be launched in October 1986, promising unprecedented high-resolution images of the universe.

14 R. A. Elson and S. M. Fall, 'Age calibration and age distribution for rich star clusters in the Large Magellanic Cloud', *The Astrophysical Journal*, 299 (1985), 211–18.

15 R. A. Elson and S. M. Fall, 'A luminosity function for star clusters in the Large Magellanic Cloud', *Publications of the Astronomical Society of the Pacific*, 97.594 (1985), 692–6.

16 On 28 January 1986 the tenth flight of the NASA Space Shuttle *Challenger* broke apart seventy-three seconds after take-off, killing all seven crew members.

17 Elson, *A Responsibility to Awe*, p. 157.

18 Aside from what she describes as 'irksome details' (for example, while male graduates were listed by only their initials, females were required also to include either 'Miss' or 'Mrs'), Elson did not experience the same sexism at Cambridge that she did at the IAS. The astronomy department was still dominated by men, but she describes it as being less 'of a bastion of male scientists'.

19 J. B. Burnell, 'A celebration of women in astronomy', *Astronomy & Geophysics*, 45.6 (2004), 10–14 (p. 13).

20 R. Elson, P. Hut, and S. Inagaki, 'Dynamical evolution of globular clusters', *Annual Review of Astronomy and Astrophysics*, 25.1 (1987), 565–601.

21 Elson, *A Responsibility to Awe*; D. Constantine, H. Lee, and B. O'Donoghue (eds), *Oxford Poets 2000: An Anthology* (Manchester: Carcanet, 2000).

22 The Radcliffe Institute for Independent Study was founded in 1960 by the Radcliffe College president Mary Ingraham Bunting, to enable college-educated women to pursue self-directed projects. In 1978 it was renamed the Bunting Institute and provided stipends and access to all of the resources of Harvard University.

23 In 1999 Radcliffe College and Harvard University officially merged, establishing the Radcliffe Institute for Advanced Study at Harvard, which removed many of the obstacles that Elson faced during her time at the Bunting Institute.

24 M. Egan, 'Hubble, trouble, toil and space rubble: the management history of an object in space', *Management & Organizational History*, 4.3 (2009), 263–80 (p. 268).

25 D. Shiga, 'Happy birthday Hubble: the telescope that almost wasn't', *New Scientist*, 206.2756 (2010), 26–7 (p. 26).

26 Elson, *A Responsibility to Awe*, p. 20.

27 *Ibid.*, p. 129.

28 *Ibid.*, p. 14.

29 Dark energy is now thought to play an even more critical role than that of dark matter in determining the fate of the universe, and the degree to which an open universe would either slow or continue to accelerate. However, investigation of this phenomenon was only in its infancy in the mid-to-late 1990s, with the phrase 'dark energy' not coined until the very end of the 1990s.

30 Elson, *A Responsibility to Awe*, p. 159.

31 R. A. Elson, 'Surface brightness profiles for five rich star clusters in the Large Magellanic Cloud', *Monthly Notices of the Royal Astronomical Society*, 256.3 (1992), 515–18; R. A. Elson, 'The structure and evolution of rich star clusters in the Large Magellanic Cloud', *The Astrophysical Journal Supplement Series*, 76 (1991), 185–214; R. E. Griffiths, K. Ratnatunga, R. Doxsey, R. Ellis, K. Glazebrook, G. Gilmore, R. Elson, D. Schade, R. Green, F. Valdes and J. Huchra, 'The Hubble Space Telescope Medium Deep Survey: status report and first results', *European Southern Observatory Conference and Workshop Proceedings*, 33 (1992), 13–20.

32 The COSTAR was removed from the Hubble Space Telescope during a servicing mission in 2009, when it was replaced by the Cosmic Origins Spectrograph; it can now be found in the Smithsonian's National Air and Space Museum in Washington DC.

33 The Hubble Space Telescope would go on to create some of the most striking images of the universe that have ever been produced, and at the time of writing (February 2019) it remains in operation.

34 R. Zimmerman, *The Universe in a Mirror: The Saga of the Hubble Space Telescope and the Visionaries Who Built It* (Princeton, NJ: Princeton University Press, 2010), p. 180.

35 R. A. Elson and D. J. Schade, 'Structural parameters of globular clusters in the Fornax cluster', *The Astrophysical Journal*, 437 (1994), 625–9; R. E. Griffiths, K. U. Ratnatunga, L. W. Neuschaefer, S. Casertano, M. Im, E. W. Wyckoff, R. S. Ellis, G. F. Gilmore, R. A. Elson, K. Glazebrook, and D. J. Schade, 'The Hubble Space Telescope Medium Deep Survey with the wide field and planetary camera. 1: methodology and results on the field near 3C 273', *The Astrophysical Journal*, 437 (1994), 67–82.

36 R. A. Elson, B. X. Santiago, and G. F. Gilmore, 'Halo stars, starbursts, and distant globular clusters: a survey of unresolved objects in the Hubble Deep Field', *New Astronomy*, 1.1 (1996), 1–16.

37 Petà Dunstan, personal communication, 2018.

38 Elson, *A Responsibility to Awe*, p. 54.

39 A cancer of the lymphatic system.

40 Constantine, Lee, and O'Donoghue (eds), *Oxford Poets 2000*, p. 113.

41 Elson, *A Responsibility to Awe*, p. 141.

42 G. M. Cragg and D. J. Newman, 'Plants as a source of anti-cancer agents', *Journal of Ethnopharmacology*, 100.1–2 (2005), 72–9 (p. 73).

43 J. Hirsch, 'An anniversary for cancer chemotherapy', *Jama*, 296.12 (2006), 1518–20 (p. 1520).

44 Elson, *A Responsibility to Awe*, p. 136. Transumanza is the Italian term for transhumance, the practice of moving livestock from one grazing ground to another in a seasonal cycle, typically to lowlands in the winter and highlands in the summer.

45 D. A. Forbes, R. A. Elson, R. E. Griffiths, R. S. Ellis, G. Gilmore, R. F. Green, J. P. Huchra, G. D. Illingworth, D. C. Koo, A. Tyson and R. A. Windhorst, 'An HR diagram for the LMC from the medium deep survey', in *Very High Angular Resolution Imaging* (Springer, Dordrecht, 1994), pp. 404–6.

46 Constantine, Lee, and O'Donoghue (eds), *Oxford Poets 2000*, p. 113.

47 Elson, *A Responsibility to Awe*, p. 143.

48 *Ibid.*, p. 77.

49 *Ibid.*, p. 61. This is the final finished poem in this volume.

50 *Ibid.*, p. 122.

51 G. Gilmore, 'Rebecca Anne Wood Elson 1960–1999', *Astronomy & Geophysics*, 41.2 (2000), 38.

52 It is perhaps fitting that the presidential address from Bell Burnell that was mentioned earlier in this chapter is entitled 'A celebration of women in astronomy',

namechecking Elson alongside Caroline Herschel, Cecilia Payne-Gaposchkin, and other luminaries in the field.

53 Angelo di Cintio, personal communication, 2018.

54 R. A. Elson, S. Sigurdsson, J. Hurley, M. B. Davies, and G. F. Gilmore, 'Discovery of a luminous white dwarf in a young star cluster in the Large Magellanic Cloud', *The Astrophysical Journal Letters*, 499.1 (1998), L53. A white dwarf is one of a class of faint stars representing the endpoint of the evolution of intermediate- and low-mass stars, such as our sun.

55 H. Gee, 'A critical moment in stellar evolution', *Nature* (1998), https://www.nature.com/news/1998/980702/full/news980702-7.html (accessed 15 February 2019).

56 D. J. Helfand and G. F. Gilmore, 'Rebecca Anne Wood Elson', *Physics Today*, 52.9 (1999), 74–5 (p. 75).

57 Elson, *A Responsibility to Awe*, p. 86.

Notes to the Epilogue

1 R. Kelly, *May Day* (Newry, Canada: Parsifal Press, 2007), p. 44.

2 R. Elson, *A Responsibility to Awe* (Manchester: Carcanet, 2001), p. 9.

3 See, for example, J. Herman, 'Medicine: the science and the art', *Medical Humanities*, 27 (2001), 42–6.

4 P. B. Shelley, *The Major Works*, ed. Z. Leader and M. O'Neill (Oxford: Oxford University Press, 2009), p. 680.

5 See, for example, E. Miller and L. Brockie, 'The disaster flood experience: older people's poetic voices of resilience', *Journal of Aging Studies*, 34 (2015), 103–12.

6 See, for example, K. F. Jack and J. Tetley, 'Using poems to explore the meaning of compassion to undergraduate nursing students', *International Practice Development Journal*, 6.1 (2016), 1–13.

7 See, for example, S. Illingworth and K. Jack, 'Rhyme and reason – using poetry to talk to underserved audiences about environmental change', *Climate Risk Management*, 19 (2018), 120–9; S. Illingworth, 'Are scientific abstracts written in poetic verse an effective representation of the underlying research?', *F1000Research*, 5 (2016), 91–119.

8 A. Cappi, 'Edgar Allan Poe's physical cosmology', *Quarterly Journal of the Royal Astronomical Society*, 35 (1994), 177–92.

9 E. A. Poe, *Eureka: An Essay on the Material and Spiritual Universe* (Auckland: The Floating Press, 2011), p. 4.

10 A. I. Tauber, *Henry David Thoreau and the Moral Agency of Knowing* (Berkeley, CA: University of California Press, 2001), p. 137.

Select bibliography

Aristotle, *Poetics*, trans. Anthony Kenny (Oxford: Oxford University Press, 2013).

Babuts, N., 'Reading poetry: metaphors as instruments of discovery', *Symposium: A Quarterly Journal in Modern Literatures*, 57.4 (2003), 211–29.

Bamforth, I., 'Applied poetry', *Parnassus: Poetry in Review*, 25.1/2 (2001), 281–307.

Beer, G., *Darwin's Plots: Evolutionary Narrative in Darwin, George Eliot and Nineteenth-Century Fiction* (Cambridge: Cambridge University Press, 2009).

Beer, G., 'Translation or transformation? The relations of literature and science', *Notes and Records of the Royal Society of London*, 44.1 (1990), 81–99.

Brown, D., *The Poetry of Victorian Scientists: Style, Science and Nonsense* (Cambridge: Cambridge University Press, 2013).

Brown, S.-A., 'Creative expression of science through poetry and other media can enrich medical and science education', *Frontiers in Neurology*, 6.3 (2015), 1–5.

Bynum, W. F., and Overy, C. (eds), *The Beast in the Mosquito: The Correspondence of Ronald Ross and Patrick Manson* (Amsterdam and Atlanta, GA: Rodopi, 1998).

Byron, G. G., Lord, *The Major Works*, ed. J. J. McGann (Oxford: Oxford University Press, 2008).

Campbell, L., and Garnett, W., *The Life of James Clerk Maxwell* (London: Macmillan, 1882).

Collier, B., and MacLachlan, J., *Charles Babbage: And the Engines of Perfection* (Oxford: Oxford University Press, 2000).

Crawford, R., *Contemporary Poetry and Contemporary Science* (Oxford: Oxford University Press, 2006).

Davy, J., *Memoirs of the Life of Sir Humphry Davy* (1836), 2 vols (Cambridge: Cambridge University Press, 2011).

Dias, P., and Hayhoe, M., *Developing Response to Poetry* (Milton Keynes: Open University Press, 1988).

Elson, R., *A Responsibility to Awe* (Manchester: Carcanet, 2001).

Finley, M., 'Fugue of the street rat: writing research poetry', *International Journal of Qualitative Studies in Education*, 16.4 (2003), 603–4.

Flood, R., McCartney, M., and Whitaker, A. (eds), *James Clerk Maxwell: Perspectives on his Life and Work* (Oxford: Oxford University Press, 2014).

Fulford, T., and Ruston, S. (eds), *The Collected Letters of Humphry Davy*, 4 vols (Oxford: Oxford University Press, 2018).

Fullmer, J. Z., *Young Humphry Davy: The Making of an Experimental Chemist* (Philadelphia, PA: American Philosophical Society, 2000).

Glazebrook, R., *James Clerk Maxwell and Modern Physics* (London: Cassell, 1896).

Hagen, M., and Skagen, M. V. (eds), *Literature and Chemistry: Elective Affinities* (Aarhus: Aarhus University Press, 2014).

Hall, M. B., *All Scientists Now: The Royal Society in the Nineteenth Century* (Cambridge: Cambridge University Press, 2002).

Hindle, M., 'Humphry Davy and William Wordsworth: a mutual influence', *Romanticism*, 18.1 (2012), 16–29.

Hindle, M., 'Nature, power, and the light of suns: the poetry of Humphry Davy', *The Charles Lamb Bulletin*, 157 (2013), 38–54.

Holmes, J., *Science in Modern Poetry: New Directions* (Liverpool: Liverpool University Press, 2012).

Holmes, R., *The Age of Wonder: How the Romantic Generation Discovered the Beauty and Terror of Science* (New York: Vintage, 2010).

Holub, M., *Supposed to Fly: A Sequence from Pilsen, Czechoslovakia*, trans. E. Osers (Hexham: Bloodaxe, 1996).

Hutchinson, I. H., 'The genius and faith of Faraday and Maxwell', *The New Atlantis*, 41 (2014), 81–99.

Kim, D.-W., *Leadership and Creativity: A History of the Cavendish Laboratory, 1871–1919* (Dordrecht: Springer Science & Business Media, 2002).

Klancher, J., *Transfiguring the Arts and Sciences: Knowledge and Cultural Institutions in the Romantic Age* (Cambridge: Cambridge University Press, 2013).

Koch, W., *Poetry and Science: Semiogenetical Twins* (Tübingen: Gunter Narr Verlag, 1983).

Lamont-Brown, R., *Humphry Davy: Life Beyond the Lamp* (Stroud: Sutton Publishing, 2004).

Lederman, L. M., and Hill, C. T., *Quantum Physics for Poets* (Amherst, MA: Prometheus Books, 2011).

Leroi, A. M., *The Lagoon: How Aristotle Invented Science* (London: Bloomsbury, 2014).

Longair, M., *Maxwell's Enduring Legacy: A Scientific History of the Cavendish Laboratory* (Cambridge: Cambridge University Press, 2016).

Mahon, B., *The Man Who Changed Everything: The Life of James Clerk Maxwell* (Hoboken, NJ: John Wiley & Sons, 2004).

Maxwell, J. C., *The Scientific Papers of James Clerk Maxwell*, ed. W. D. Niven (Cambridge: Cambridge University Press, 1890).

McCarty, V. M., 'Poetry, science and truth: the case of "poet-scientists" Miroslav Holub and David Morley', *Interdisciplinary Science Reviews*, 39.1 (2014), 33–46.

Mégroz, R. L., *Ronald Ross, Discoverer and Creator* (London: Allen & Unwin, 1931).

Middleton, P., *Physics Envy: American Poetry and Science in the Cold War and After* (Chicago: University of Chicago Press, 2015).

Midgley, M., *Science and Poetry* (Abingdon: Routledge, 2006).

Nye, E., and Gibson, M., *Ronald Ross: Malariologist and Polymath: A Biography* (Dordrecht: Springer, 1997).

Padua, S., *The Thrilling Adventures of Lovelace and Babbage: The (Mostly) True Story of the First Computer* (London: Penguin, 2016).

Paris, J. A., *The Life of Sir Humphry Davy* (1831), 2 vols (Cambridge: Cambridge University Press, 2011).

Poe, E. A., *The Complete Tales & Poems of Edgar Allan Poe* (New York: Race Point Publishing, 2014).

Priestman, M., *The Poetry of Erasmus Darwin: Enlightened Spaces, Romantic Times* (Abingdon: Routledge, 2016).

Quinn, J., *Between Two Fires: Transnationalism and Cold War Poetry* (Oxford: Oxford University Press, 2015).

Riordan, M., and Turney, J. (eds), *A Quark for Mister Mark: 101 Poems about Science* (London: Faber, 2000).

Ross, R., *Memoirs: With a Full Account of the Great Malaria Problem and its Solution* (London: John Murray, 1923).

Shelley, P. B., *The Major Works*, ed. Z. Leader and M. O'Neill (Oxford: Oxford University Press, 2009).

Toole, B. A., *Ada, the Enchantress of Numbers: A Selection from the Letters of Lord Byron's Daughter and Her Description of the First Computer* (Sausalito, CA: Strawberry Press, 1992).

Zimmerman, R., *The Universe in a Mirror: The Saga of the Hubble Space Telescope and the Visionaries Who Built It* (Princeton, NJ: Princeton University Press, 2010).

Index

Campbell, Lewis 67, 70, 73, 75, 86–8
Campbell, William 74
Camus, Albert 140
Čapek, Karel 125
Cardew, Cornelius 12–13
cathodic protection 38–9
Cavendish, Henry 78–9
Cavendish, William 78
Cavendish Laboratory 78–9, 84
censorship 136–8, 147–9
Challenger disaster (1986) 158,
161
chemical reactions 22
Clanny, William Reid 30–2
Clare, John 5–6
Clement, Joseph 45
clerihews 41–2
Clerk Maxwell, James, *see* Maxwell
Cohen, Leonard 158
Coleridge, Samuel Taylor 9, 18–23, 26,
28, 93, 158
'Kubla Khan' 20
communist regimes 136
complementarity of the worlds of science
and poetry 178
Corbière, Tristan 123
Coryton, George 12–13, 42
Coulomb, Charles-Augustin de 76
Cousteau, Jacques 153
Crick, Francis 79
Croker, John Wilson 35
Cronstedt, Axel Fredrik 96
Čulík, Jan 135
Czechoslovakia
overthrow of communist regime (1989)
138
Warsaw Pact invasion of (1968) 130–1,
134

Daniels, Charles Wilberforce 110
Dante Alighieri 158
dark matter 163–5
Darwin, Charles 80–1
Darwin, Erasmus 8–9, 18
Davy (née Miller), Grace 11, 17
Davy, Humphry 9, 11–42, 56, 96, 103,
152, 175
as an author 39–40

concentration on the pursuit of science
25
death and burial 41
illnesses and injuries suffered by 28–30,
39–40
influence of 41–2
invention of the safety lamp 29–32
knighthood 29
as a lecturer 25–7
notebooks kept by 14–16
personal philosophy 22–3
poems by 15–16, 20–4, 28, 32–7, 40–2
as president of the Royal Society 35–6,
39
scientific discoveries made by 27–30,
39, 42
works
'On breathing the Nitrous Oxide' 19
'On the Death of Lord Byron' 34–5
'Sons of Genius' 15–16, 18
'Thoughts after the ingratitude ...'
32–3
'Written after Recovery from a
Dangerous Illness' 22–3
Davy (née Apreece), Jane 29, 34
Davy, John 12–14, 37
Delphi, Temple of 115
De Morgan, Augustus 45
Denis, John 17
Descartes, René 69
Dewar, Daniel 75
di Cintio, Angelo 166–7, 170–2
Dickens, Charles 45
Douglas, James 67–8
Doyle, Arthur Conan 116
drug-taking, recreational 19
Dubček, Alexander 130
Dunkin, Robert 16–17
Dunstan, Petà 157–8, 166
Durham, Earl of, *see* Lambton

Edinburgh Academy 66–70
Edinburgh University 69–71, 76–8
Einstein, Albert 124
electrolysis 27–8
electromagnetism, research into 76–7,
86, 88
Elson, John 151–3